Name ______________________________ Class ______

Skills Worksheet

Directed Reading

Section: How Organisms Interact in Communities

In the space provided, explain how the terms in each pair differ in meaning.

1. coevolution, secondary compounds

____

____

____

2. predation, parasitism

____

____

____

Complete each statement by writing the correct term or phrase in the space provided.

3. The evolution of flowering plants and the insects that transport their male gametes is an example of ______________________ .

4. The interaction between mosquitoes and human beings is called ______________________ .

5. The relationship between small insects called aphids and ants is called ______________________ .

6. The relationship between certain small tropical fishes and sea anemones is an example of ______________________ .

7. When two or more species live together in a close, long-term relationship, it is called ______________________ .

8. A symbiotic relationship in which both participating species benefit is called ______________________ .

9. A symbiotic relationship in which one species is neither harmed nor helped is called ______________________ .

Copyright © by Holt, Rinehart and Winston. All rights reserved.

Name ______________________ Class ______________ Date ____________

Skills Worksheet

Directed Reading

Section: How Competition Shapes Communities

In the space provided, write the letter of the description that best matches the term or phrase.

______ **1.** competition

______ **2.** niche

______ **3.** fundamental niche

______ **4.** realized niche

______ **5.** competitive exclusion

______ **6.** biodiversity

______ **7.** species richness

______ **8.** species diversity

______ **9.** productivity

a. the functional role of a particular species in an ecosystem

b. the entire range of conditions an organism is potentially able to occupy

c. biological interaction in which two species use the same resources

d. the part of a fundamental niche that a species actually occupies

e. the variety of living organisms living in a community

f. the relative numbers of each of the species living in a community

g. the amount of plant material produced in a plot of land

h. elimination of a competitive species

i. the number of different species in a community

Read each question, and write your answer in the space provided.

10. What are the different niches of *Chthamalus stellatus* and *Semibalanus balanoides*?

____

____

____

11. What happened in Connell's experiment with *Chthamalus stellatus* and *Semibalanus balanoides*?

____

____

____

Copyright © by Holt, Rinehart and Winston. All rights reserved.

12. Why was *Chthamalus* unable to compete with *Semibalanus* at the lower depths?

____

____

13. Why was *Semibalanus* unable to survive in shallow water?

____

____

Complete each statement by underlining the correct term or phrase in the brackets.

14. In the experiments by G. F. Gause, *Paramecium* fed on [bacteria / culture].

15. The smaller species of *Paramecium* in the first experiment was [more / less] resistant to bacterial waste products.

16. The process by which the smaller species of *Paramecium* drove the larger species to extinction is called [survival of the fittest / competitive exclusion].

17. In the second experiment, *Paramecium caudatum* [coexisted with / eliminated] *Paramecium bursaris.*

18. *P. caudatum* and *P. bursaris* had [different niches / the same niche].

Read each question, and write your answer in the space provided.

19. In the studies of Robert Paine, why did eliminating sea stars cause the number of species to decrease?

____

____

____

20. Describe two aspects of a community that biodiversity measures.

____

____

21. What is the relationship between biodiversity and productivity?

____

____

Copyright © by Holt, Rinehart and Winston. All rights reserved.

Name ______________________ Class ______________ Date ______________

Skills Worksheet

Directed Reading

Section: Major Biological Communities

In the space provided, write the letter of the description that best matches the term or phrase.

_____ **1.** tropical rain forests

_____ **2.** deserts

_____ **3.** savannas

_____ **4.** temperate deciduous forests

_____ **5.** temperate grasslands

_____ **6.** taiga

_____ **7.** tundra

a. water is unavailable for most of the year because it is frozen

b. northern forests of coniferous trees

c. warm summers, cold winters, and sufficient precipitation

d. another name for prairie; contains deep and fertile soil

e. has the greatest number of species; has a very infertile soil

f. landscape has widely spaced trees; seasonal drought

g. vegetation is very sparse

Complete each statement by writing the correct term or phrase in the space provided.

8. The prevailing weather conditions in any given area is called

______________________ .

9. The growing season of plants is primarily influenced by ______________________ .

10. The moisture-holding ability of air ______________________ when it is warmed and ______________________ when it is cooled.

11. A major biological community that occurs over a large area of land is called a(n) ______________________ .

12. In general, temperature and moisture ______________________ as distance from the equator ______________________ .

13. The shallow area of ponds and lakes, near the shore, is called the

______________________ ______________________ .

14. The ______________________ ______________________ of lakes and ponds is away from the shore but close to the surface.

Copyright © by Holt, Rinehart and Winston. All rights reserved.

15. The ______________________ ______________________ is a deep-water zone that is below the limits of effective light.

16. Nearly three-fourths of the Earth's surface is covered by ______________________ .

17. Small organisms that drift in the upper waters of the ocean are called ______________________ .

Copyright © by Holt, Rinehart and Winston. All rights reserved.

Name ______________________ Class ______________ Date ____________

Skills Worksheet

Active Reading

Section: How Organisms Interact in Communities

Read the passage below. Then answer the questions that follow.

In **symbiosis,** two or more species live together in a close, long-term association. Symbiotic relationships can be beneficial to both organisms or may benefit one organism and leave the other harmed or unaffected. **Parasitism** is one type of symbiotic relationship that is detrimental to, or harms, the host organism. In this relationship, one organism feeds on and usually lives in another, typically larger, organism. **Mutualism** is a symbiotic relationship in which both participating species benefit. A well-known instance of mutualism involves ants and aphids. The ants feed on fluid the aphids secrete, and in exchange, the ants protect the aphids from insect predators. A third form of symbiosis is **commensalism,** a symbiotic relationship in which one species benefits and the other is neither harmed nor helped. Among the best-known examples of commensalism are the feeding and protection relationships between certain small tropical fishes and sea anemones, marine animals that have stinging tentacles.

SKILL: READING EFFECTIVELY

Write *P* if the phrase describes parasitism, *M* if it describes mutualism, or *C* if it describes commensalism. For each question, some choices may be used more than once.

______ **1.** exists between certain tropical fish and sea anemones

______ **2.** type of symbiotic relationship

______ **3.** the host organism is harmed

______ **4.** one species is neither harmed nor helped

______ **5.** at least one species benefits

In the space provided, write the letter of the phrase that best completes the statement.

______ **6.** Mutualism is a symbiotic relationship in which
a. both species are harmed.
b. neither species benefits.
c. one species is harmed.
d. both species benefit.

Copyright © by Holt, Rinehart and Winston. All rights reserved.

Name ______________________ Class ______________ Date ______________

Skills Worksheet

Active Reading

Section: How Competition Shapes Communities

Read the passage below. Then answer the questions that follow.

A key investigation carried out in the early 1990s by David Tilman of the University of Minnesota illustrates the relationship between **biodiversity,** which is the variety of living organisms present in a community, and productivity. Tilman, along with co-workers and students, tended 147 experimental plots in a Minnesota prairie. Each plot contained a mix of up to 24 native prairie plant species. The biologist monitored the plots, measuring how much growth was occurring. Tilman found that the more species a plot had, the greater the amount of plant material produced in that plot. Tilman's experiments clearly demonstrated that increased biodiversity leads to greater productivity. Tilman also found that the plots with greater numbers of species recovered more fully from a major drought. Thus, the biologically diverse plots were also more stable than the plots with fewer species.

SKILL: READING EFFECTIVELY

Read each question, and write your answer in the space provided.

1. What sentence expresses the main idea of the passage? What is the main idea?

____

____

____

2. What was the location and focus of Tilman's investigation?

____

____

In the space provided, write the letter of the phrase that best completes the statement.

______ **3.** According to the passage, the greater the variety of living organisms present in a community, the greater the

a. size of the community.
b. stability of the community.
c. amount of plant material used by the community.
d. Both (a) and (b)

Copyright © by Holt, Rinehart and Winston. All rights reserved.

Name ______________________ Class ______________ Date ____________

Skills Worksheet

Active Reading

Section: Major Biological Communities

Read the passage below. Then answer the questions that follow.

A major biological community that occurs over a large area of land is called a **biome.** A biome's structure and appearance are similar throughout its geographic distribution. While there are different ways of classifying biomes, the classification system used here recognizes seven widely occurring biomes: tropical rain forest, desert, savanna, temperate deciduous forest, temperate grassland, taiga, and tundra. These biomes differ greatly from one another because they have developed in regions with very different climates.

In general, temperature and available moisture decrease as latitude (distance from the equator) increases. They also decrease as elevation (height above sea level) increases.

SKILL: READING EFFECTIVELY

Read each question, and write your answer in the space provided.

1. What key term is defined in the opening sentence of this passage?

____

2. According to the classification system used by this text, what seven biomes occur on Earth?

____

____

____

3. What happens to temperature and available moisture as elevation increases?

____

____

In the space provided, write the letter of the phrase that best answers the question.

______ **4.** What occurs as latitude decreases?

a. Temperature increases.
b. Available moisture increases.
c. Elevation decreases.
d. Both (a) and (b)

Copyright © by Holt, Rinehart and Winston. All rights reserved.

Name ______________________ Class ______________ Date ____________

Skills Worksheet

Vocabulary Review

In the space provided, write the letter of the description that best matches the term or phrase.

_____ **1.** coevolution

_____ **2.** predation

_____ **3.** parasitism

_____ **4.** secondary compound

_____ **5.** symbiosis

_____ **6.** mutualism

_____ **7.** commensalism

_____ **8.** competition

_____ **9.** niche

_____ **10.** fundamental niche

_____ **11.** realized niche

_____ **12.** competitive exclusion

_____ **13.** biodiversity

a. defensive chemical used by plants

b. a relationship in which both participating species benefit

c. the entire range of conditions an organism is potentially able to occupy

d. when two species use the same resource

e. back-and-forth evolutionary adjustments between interacting members of an ecosystem

f. two or more species living together in a close, long-term relationship

g. the fundamental role of a species in an ecosystem

h. one organism feeds on and usually lives on or in another larger organism

i. the elimination of a competing species

j. the part of its fundamental niche that a species occupies

k. a relationship in which one species benefits and the other is neither harmed nor helped

l. the variety of living organisms in a community

m. the act of one organism feeding on another

Complete each statement by writing the correct term or phrase in the space provided.

14. The prevailing weather conditions in any given area are called the ____________________ .

15. A(n) ____________________ is a major biological community that occurs over a large area of land.

16. The ____________________ ____________________ is a shallow zone near the shore.

Copyright © by Holt, Rinehart and Winston. All rights reserved.

17. The ________________ ________________ is away from the shore but close to the surface.

18. The ________________ ________________ is a deep-water zone below the limits of effective light penetration.

19. Small organisms that drift in the upper waters of the ocean are called ________________ .

Copyright © by Holt, Rinehart and Winston. All rights reserved.

Name ______________________________ Class ______________ Date ____________

Skills Worksheet

Science Skills

Interpreting Maps/Interpreting Tables

Use the map below, which shows the major terrestrial biomes of North America and Central America, to complete items 1–9 below.

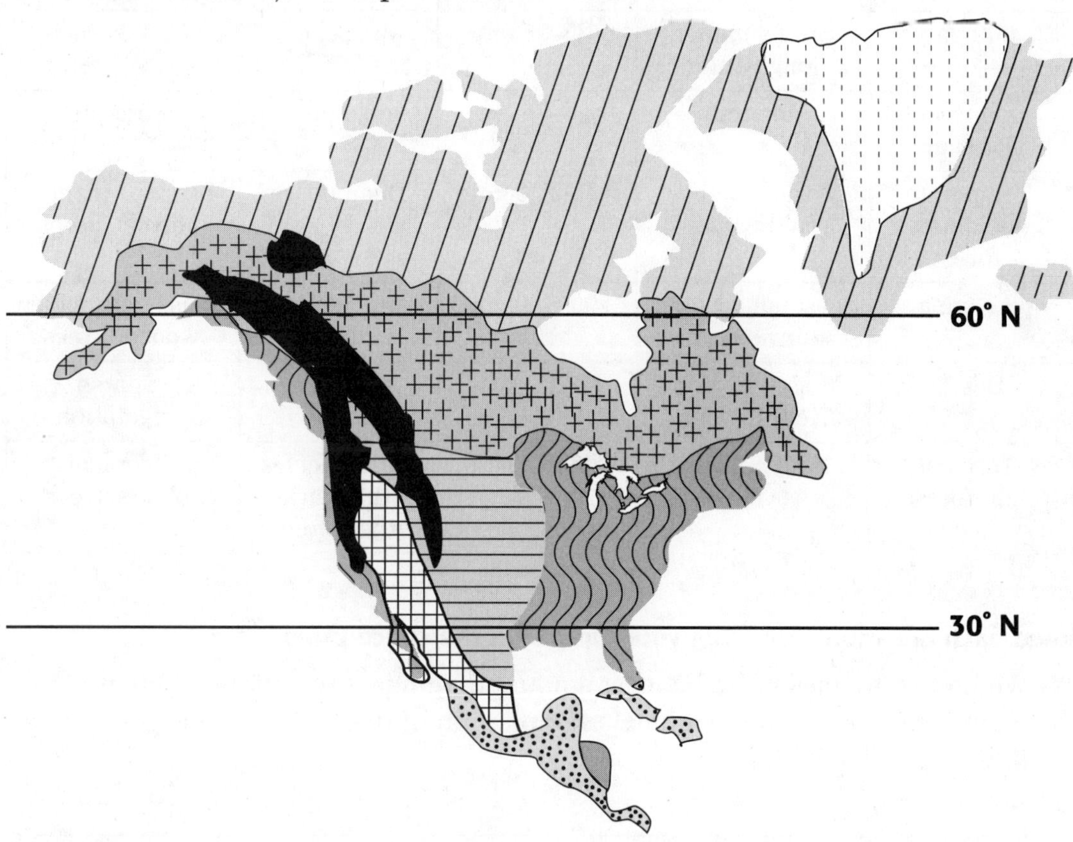

In the space provided, write the correct name of each biome next to its key.

1. ____
2. ____
3. ____
4. ____
5. ____
6. ____
7. ____
8. ____
9. ____

Copyright © by Holt, Rinehart and Winston. All rights reserved.

Use the table below to answer questions 10 and 11.

Biome	Climate	Annual precipitation	Animal life	Vegetation
Tundra	Brief summer, long winter	<25 cm	Caribou, ducks	Dwarf willows
Taiga	Brief summer, long winter	35–75 cm	Moose, elk	Firs, spruce
Temperate grassland	Moderate	25–75 cm	Bison, antelope	Grasses
Temperate deciduous forest	Warm summer, cold winter	75–250 cm	Deer, bears	Birches, maples, shrubs, herbs
Savanna	Seasonal drought, rainy season	90–150 cm	Large herds of grazing animals	Grass with widely spaced trees
Desert	Moisture varies year to year	<25 cm	Tortoises, jackrabbits	Sparse vegetation
Tropical rain forest	Rain falls evenly	200–450 cm	More species than any other biome	Tropical plants, trees

Read each question, and write your answer in the space provided.

10. Which two biomes have the least amount of annual precipitation? What is the relationship between the annual precipitation of these biomes and the vegetation they can support?

____

____

____

____

11. How are the temperature and available moisture of a biome related to the biome's distance from the equator?

____

____

____

____

Copyright © by Holt, Rinehart and Winston. All rights reserved.

Name ______________________ Class ______________ Date ____________

Skills Worksheet

Concept Mapping

Using the terms and phrases provided below, complete the concept map showing the characteristics of biological communities.

biomes	mutualism	predation
competition	niche	symbiosis
fundamental niche	parasitism	

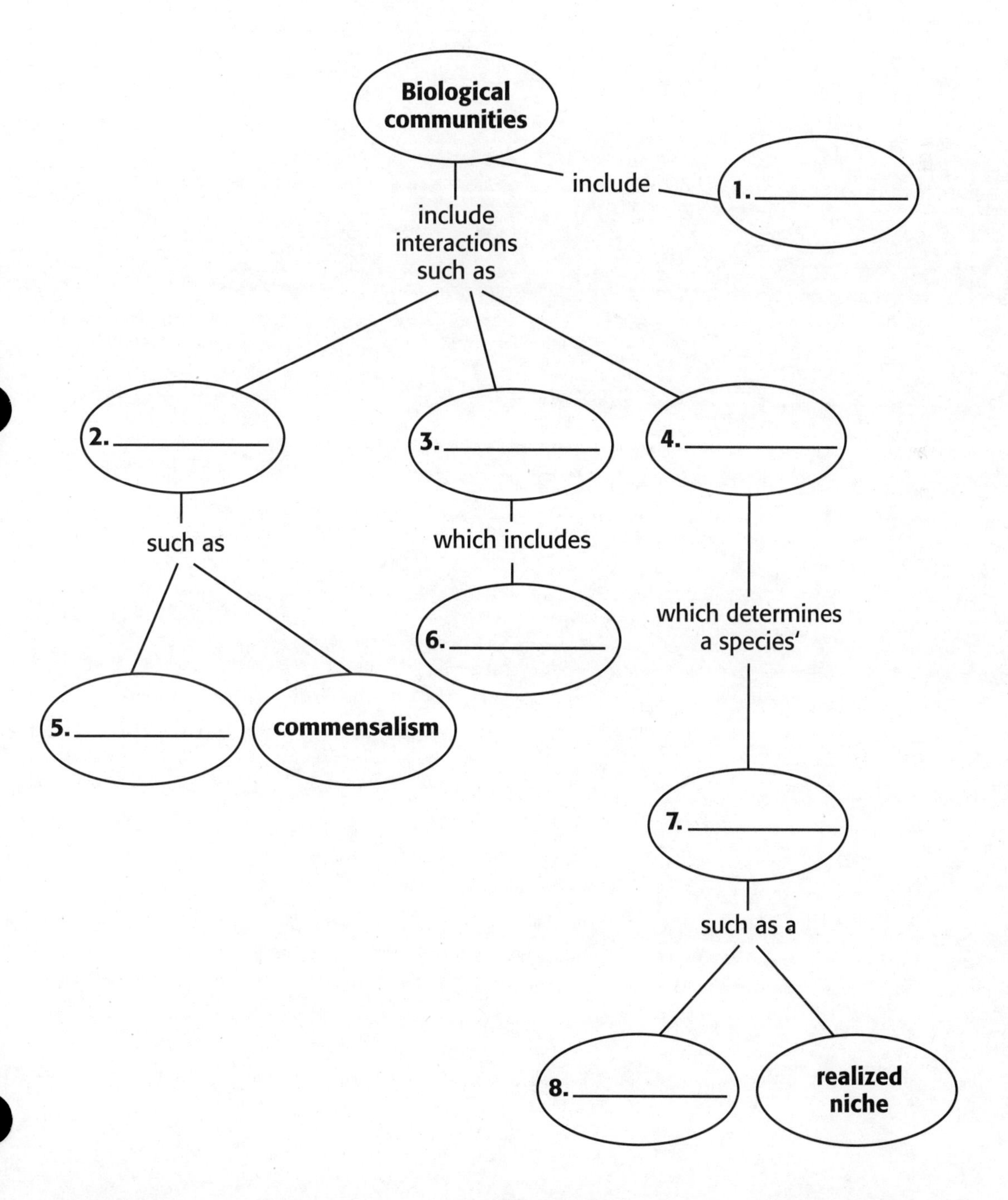

Copyright © by Holt, Rinehart and Winston. All rights reserved.

Name ______________________ Class ______________ Date ______________

Skills Worksheet

Critical Thinking

Work-Alikes

In the space provided, write the letter of the term or phrase that best describes how each numbered item functions.

_____ **1.** mutualism
_____ **2.** plant secondary compounds
_____ **3.** niche
_____ **4.** competitive exclusion

a. cooperative neighbors
b. a store goes out of business because of the success of another store
c. chemical warfare
d. job

Cause and Effect

In the space provided, write the letter of the term or phrase that best matches each cause or effect given below.

Cause	Effect
5. coevolution	______________
6. ______________	clown fish have a safe place to live
7. increased biodiversity in an ecosystem	______________
8. air is warmed	______________
9. ______________	temperate deciduous forest

a. moisture-holding capacity of air increases
b. mild climate and plentiful rain
c. flowering plants and pollinators have a long evolutionary history together
d. greater productivity
e. commensalism

Trade-offs

In the space provided, write the letter of the bad news item that best matches each numbered good news item below.

Good News

_____ **10.** Parasites do not usually kill their prey.
_____ **11.** Many plants are protected from herbivores by defensive chemicals called secondary compounds.
_____ **12.** An ocean shallow-water zone is inhabited by a large number of species.
_____ **13.** Tropical rain forests have high primary productivity.

Bad News

a. Their soils are very infertile.
b. Some animals can break them down.
c. They spread their offspring from the host.
d. It is a relatively small area.

Copyright © by Holt, Rinehart and Winston. All rights reserved.

Linkages

In the spaces provided, write the letters of the two terms or phrases that are linked together by the term or phrase in the middle. The choices can be placed in any order.

14. _____ cabbage butterflies can break down mustard oils _____

15. _____ niches did not overlap _____

16. _____ sea stars were removed from an experimental plot _____

17. _____ savanna _____

18. _____ tundra _____

19. _____ limnetic zone _____

a. cabbage butterflies feed on mustard family plants

b. the number of different prey species dropped

c. taiga

d. *P. caudatum* and *P. bursaria* are introduced into the same culture tube

e. profundal zone

f. littoral zone

g. tropical rain forest

h. sea stars are fierce predators of marine animals

i. permanent ice

j. both species survived

k. desert

l. the mustard plant family produces mustard oils

Anal ogies

An analogy is a relationship between two pairs of terms or phrases written as a : b :: c : d. The symbol : is read as "is to," and the symbol :: is read as "as." In the space provided, write the letter of the pair of terms or phrases that best completes the analogy shown.

_____ **20.** lions hunting zebras : predation ::
- **a.** animals fighting over nesting areas : predation
- **b.** lions fighting hyenas for food : competition
- **c.** plants growing over each other : predation
- **d.** organisms living inside other organisms : competition

_____ **21.** species richness : biodiversity ::
- **a.** realized niche : fundamental niche
- **b.** fundamental niche : realized niche
- **c.** large supply of resource : short supply
- **d.** large supply of resource : competition

_____ **22.** temperature and moisture decreases : latitude increases ::
- **a.** temperature and moisture : mountain height decreases
- **b.** latitude : latitude increases
- **c.** temperature and moisture : latitude decreases
- **d.** temperature and moisture decreases : altitude increases

_____ **23.** prairie : temperate grassland ::
- **a.** tundra : prairie
- **b.** savanna : desert
- **c.** desert : savanna
- **d.** northern coniferous forest : taiga

Copyright © by Holt, Rinehart and Winston. All rights reserved.

Name ______________________ Class ______________ Date ____________

Skills Worksheet

Test Prep Pretest

In the space provided, write the letter of the term or phrase that best completes each statement or best answers each question.

______ **1.** What form of interaction is taking place when a shark devours a seal?
a. commensalism
b. mutualism
c. predation
d. parasitism

______ **2.** When lions and hyenas fight over a dead zebra, their interaction is called
a. mutualism.
b. competition.
c. commensalism.
d. parasitism.

______ **3.** Mutualism and commensalism are two types of
a. symbiosis.
b. competition.
c. parasitism.
d. predation.

______ **4.** In the face of competition, an organism may occupy only part of its fundamental niche. That part is called its
a. biome.
b. community.
c. realized niche.
d. ecosystem.

______ **5.** Which of the following is NOT part of a freshwater habitat?
a. profundal zone
b. tidal zone
c. littoral zone
d. limnetic zone

______ **6.** The most important elements of climate are
a. temperature and weather.
b. temperature and moisture.
c. moisture and sun.
d. rainfall and snowfall.

______ **7.** The greater a community's biodiversity is, the greater is its
a. productivity and stability.
b. drought-tolerance.
c. degree of competition.
d. Both (a) and (b)

______ **8.** A major biological community that occurs over a large area of land is called a
a. biome.
b. profundal zone.
c. niche.
d. population.

______ **9.** Biomes characterized by high annual rainfalls are generally located at
a. high elevations.
b. low latitudes.
c. high latitudes.
d. Both (a) and (c)

______ **10.** Herds of grazing mammals are found in
a. the taiga.
b. the tropical forest.
c. the savanna.
d. the desert.

Copyright © by Holt, Rinehart and Winston. All rights reserved.

Complete each statement by writing the correct term or phrase in the space provided.

11. A characteristic of ______________________ is that they often do not kill their prey because they depend on the prey for food, a place to live, and a means to transmit their offspring.

12. Virtually all plants contain defensive chemicals called ______________________ ______________________.

13. Mild climate and annual precipitation of 75–250 cm favor the growth of the type of biome called ______________________ ______________________ ______________________.

14. The entire range of conditions an organism can tolerate is its ______________________ ______________________.

15. Back-and-forth evolutionary adjustments between interacting members of an ecosystem are called ______________________.

16. When sea stars were kept out of experimental plots in the coastal community studied by Robert Paine, the number of species in the ecosystem ______________________.

17. Fewer than 25 cm of precipitation per year falls in two of the world's biomes—the desert and the ______________________.

18. Latitude is a measure of distance from the ______________________.

19. A prairie is a biome called ______________________ ______________________.

Read each question, and write your answer in the space provided.

20. Why is parasitism considered a special case of predation?

____

____

____

____

Copyright © by Holt, Rinehart and Winston. All rights reserved.

21. Explain how the larvae of the cabbage butterfly have overcome the mustard plant's defenses.

____

____

____

____

22. Explain how predation, competition, and biodiversity are related.

____

____

____

____

____

____

23. Explain how three species of warblers that consume insects in spruce trees can occupy the same forest without violating Gause's principle.

____

____

____

____

24. Where are the world's most abundant fishing grounds located? Explain.

____

____

____

____

Copyright © by Holt, Rinehart and Winston. All rights reserved.

Question 25 refers to the figure below, which shows the results of Gause's experiments with paramecia.

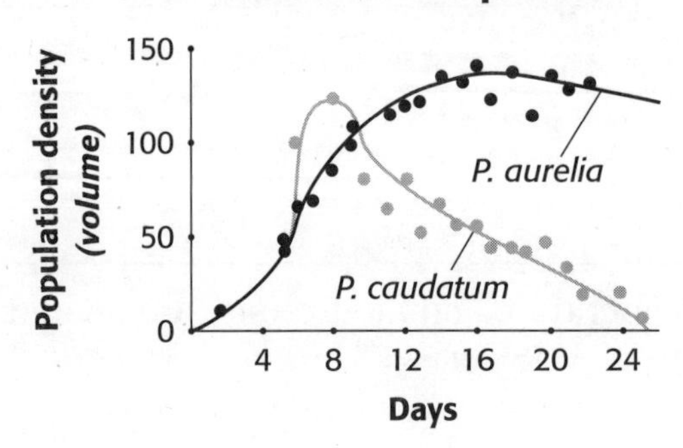

25. What principle does this graph illustrate?

____

____

Copyright © by Holt, Rinehart and Winston. All rights reserved.

Name ______________________ Class ______________ Date ____________

Assessment

Quiz

Section: How Organisms Interact in Communities

In the space provided, write the letter of the term or phrase that best completes each statement or best answers each question.

______ **1.** An adaptation in flowers and plants that promotes the efficient dispersal of pollen by insects and other animals may have arisen through
a. predation.
b. commensalism.
c. coevolution.
d. parasitism.

______ **2.** In parasitism, the host
a. is killed by the parasite.
b. usually kills the parasite.
c. is benefited by the parasite.
d. often transmits the parasite's offspring to new hosts.

______ **3.** Plants often produce secondary compounds that protect them from
a. predation.
b. parasitism.
c. mustard oils.
d. symbiotic relationships.

______ **4.** How are the larvae of cabbage butterflies able to feed on plants that have defensive chemicals?
a. They feed on other plants that counter the effects of the defensive chemicals.
b. They have adaptations that break down the secondary compounds of the plant.
c. The larvae can only feed on the plants at certain times of the year.
d. None of the above

______ **5.** Which pair of organisms exists in a commensal relationship?
a. bear and fish
b. ant and aphid
c. clown fish and sea anemone
d. dog and flea

In the space provided, write the letter of the description that best matches the term or phrase.

______ **6.** coevolution
______ **7.** predation
______ **8.** parasitism
______ **9.** mutualism
______ **10.** commensalism

a. a symbiotic relationship where one benefits and the other is neither harmed nor helped
b. one organism feeds on and usually lives on or in another larger organism
c. evolutionary adjustments between interacting members of an ecosystem
d. the act of one organism killing and eating another for food
e. a symbiotic relationship in which both members benefit

Copyright © by Holt, Rinehart and Winston. All rights reserved.

Name ______________________________ Class ______________ Date ____________

Assessment

Quiz

Section: How Competition Shapes Communities

In the space provided, write the letter of the term or phrase that best completes each statement or best answers each question.

______ **1.** An organism's niche includes
- **a.** what it eats.
- **b.** where it eats.
- **c.** how it reproduces.
- **d.** All of the above

______ **2.** When two species compete for limited resources, competitive exclusion
- **a.** is sure to take place.
- **b.** is not possible.
- **c.** will take place unless the species divide or find different resources.
- **d.** will cause both species to become extinct.

______ **3.** *Chthamalus stellatus* can live in both shallow water and deep water on a rocky coast. This is the barnacle's
- **a.** fundamental niche.
- **b.** realized niche.
- **c.** community.
- **d.** habitat.

______ **4.** What is the principle that enables five species of warbler to feed in the same tree without competing?
- **a.** commensalism
- **b.** resource partitioning
- **c.** mutualism
- **d.** competitive exclusion

______ **5.** Higher productivity, a more stable ecosystem, and reduced competition are all benefits of
- **a.** a high biodiversity.
- **b.** a low rate of predation.
- **c.** a high rate of predation.
- **d.** a low biodiversity

In the space provided, write the letter of the description that best matches the term or phrase.

______ **6.** competition

______ **7.** realized niche

______ **8.** fundamental niche

______ **9.** competitive exclusion

______ **10.** biodiversity

- **a.** the entire range of resources in an ecosystem that an organism can potentially occupy
- **b.** the biological interaction that occurs when two species use the same resource
- **c.** the variety of living organisms present in a community
- **d.** the part of its niche that a species actually occupies
- **e.** the elimination of a species due to competition

Copyright © by Holt, Rinehart and Winston. All rights reserved.

Name ______________________ Class ______________ Date ____________

Assessment

Quiz

Section: Major Biological Communities

In the space provided, write the letter of the term or phrase that best completes each statement or best answers each question.

______ **1.** Deserts and tundra have similar
- **a.** climates.
- **b.** animal inhabitants.
- **c.** annual rainfalls.
- **d.** latitudes.

______ **2.** The dry grasslands in tropical areas make up the biome called
- **a.** taiga.
- **b.** tundra.
- **c.** desert.
- **d.** savanna.

______ **3.** Forty percent of all photosynthesis on Earth is accomplished by
- **a.** photosynthetic plants.
- **b.** fungi.
- **c.** fish larvae.
- **d.** photosynthetic plankton.

______ **4.** One of the most important elements of climate is which of the following?
- **a.** temperature
- **b.** elevation
- **c.** latitude
- **d.** biome

______ **5.** Foxes, lemmings, owls, and caribou are among the vertebrate inhabitants of the
- **a.** taiga.
- **b.** tundra.
- **c.** intertidal zone.
- **d.** desert.

In the space provided, write the letter of the description that best matches the term or phrase.

______ **6.** climate

______ **7.** biome

______ **8.** profundal zone

______ **9.** limnetic zone

______ **10.** plankton

- **a.** area of fresh water that is far from shore but close to the surface
- **b.** food for some marine organisms that is composed of bacteria, algae, fish larvae, and many small invertebrates
- **c.** a major biological community that occurs over a large area of land
- **d.** freshwater zone in deep water below the limits of light penetration
- **e.** determines what kind of organisms live in a given environment

Copyright © by Holt, Rinehart and Winston. All rights reserved.

Name ______________________ Class ______________ Date ____________

Assessment

Chapter Test

Biological Communities

In the space provided, write the letter of the term or phrase that best completes each statement or best answers each question.

_____ **1.** The reciprocal evolution of flowers and the insects that feed on them is known as

a. parasitism.
b. secondary succession.
c. coevolution.
d. stability.

_____ **2.** A tick feeding on a human is an example of

a. parasitism.
b. mutualism.
c. symbiosis.
d. predation.

_____ **3.** Over time, selection pressure from predators will cause prey species to evolve

a. into parasites.
b. into a new niche.
c. secondary compounds.
d. ways to avoid predation.

_____ **4.** Tropical rain forests are found close to the equator because

a. rainfall is highest at lower latitudes.
b. soil is most fertile at lower latitudes.
c. rainfall varies from season to season near the equator.
d. rainfall averages 80 cm per year at the equator.

_____ **5.** Which of the following determines what organisms can live in a particular biological community?

a. climate
b. rainfall
c. temperature
d. All of the above

_____ **6.** Compared with a forest that contains 25 different plant species, a forest that contains 55 different plant species is

a. larger and more stable.
b. more productive and more stable.
c. older and more productive.
d. older and larger.

Copyright © by Holt, Rinehart and Winston. All rights reserved.

______ **7.** An ecologist studying an ocean ecosystem performed an experiment in which predatory sea stars were removed from the ecosystem. After the removal of the sea stars,
a. the ecosystem became more diverse.
b. the size of the ecosystem was reduced.
c. food webs in the ecosystem became more complex.
d. the number of species in the ecosystem was reduced.

Questions 8–10 refer to the figures below, which illustrate experiments performed with two species of barnacles that live in the same area.

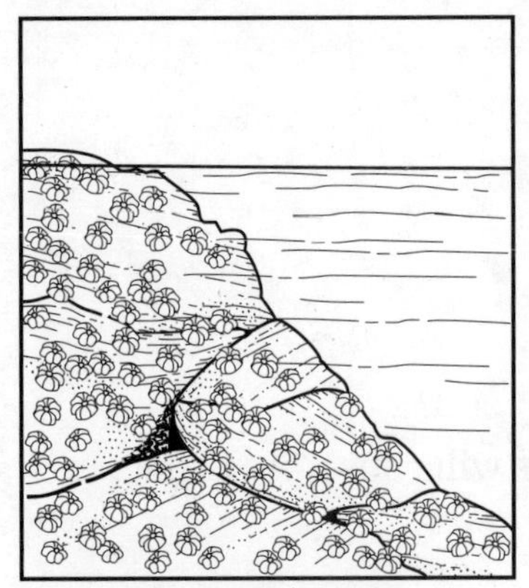

A. The barnacle *Chthamalus stellatus* can live in both shallow and deep water on a rocky coast.

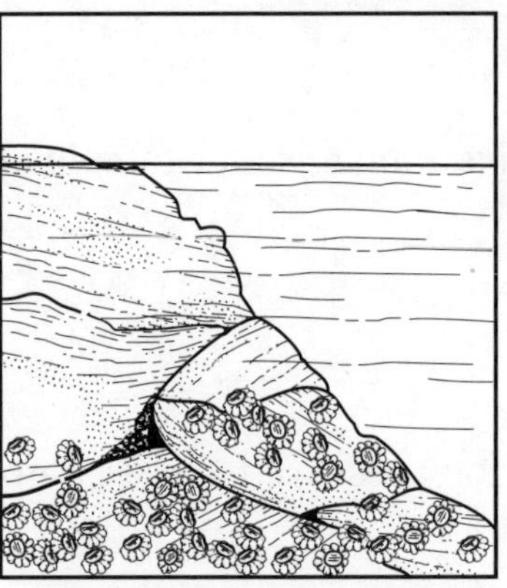

B. The barnacle *Semibalanus balanoides* lives mostly in deep water

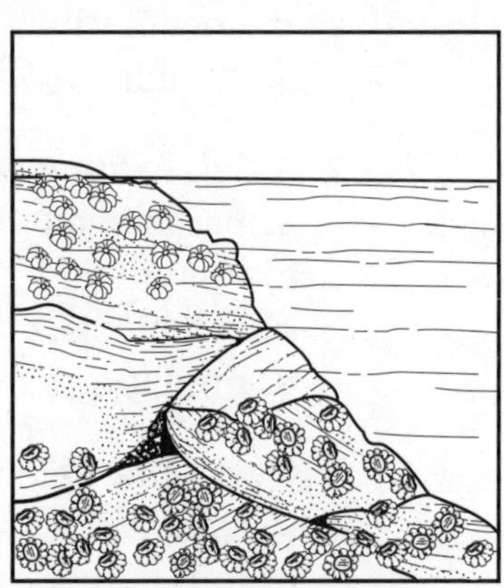

C. When the two barnacles live together, *Chthamalus* is restricted to shallow water.

______ **8.** Figure B indicates that *Semibalanus balanoides* lives mostly in deep water. Deep water is this barnacle's
a. competitive niche.
b. realized niche.
c. fundamental niche.
d. exclusive niche.

______ **9.** Figure C indicates that when the two barnacles live together, *Chthamalus* is restricted to shallow water. Shallow water is its
a. competitive niche.
b. realized niche.
c. fundamental niche.
d. exclusive niche.

______ **10.** Because the two species of barnacles attempt to use the same resources, they are
a. parasites on each other.
b. in competition with each other.
c. have a mutualistic relationship.
d. have a symbiotic relationship.

Copyright © by Holt, Rinehart and Winston. All rights reserved.

In the space provided, write the letter of the description that best matches the term or phrase.

______ **11.** tropical rain forest

______ **12.** commensalism

______ **13.** taiga

______ **14.** littoral zone

______ **15.** limnetic zone

______ **16.** mutualism

______ **17.** biodiversity

______ **18.** latitude

______ **19.** elevation

______ **20.** tundra

a. predominated by aquatic plants in freshwater communities

b. a symbiotic relationship in which both participating species benefit

c. benefits ecosystems by reducing competition and promoting stability and productivity

d. temperature and available moisture decrease as this increases

e. a relationship in which one species benefits and the other is neither harmed nor helped

f. biome predominated by coniferous trees

g. vertebrate inhabitants include foxes, lemmings, and caribou

h. the biome with the greatest amount of rainfall

i. predominated by floating algae, zooplankton, and fish

j. a measure of height above sea level

Copyright © by Holt, Rinehart and Winston. All rights reserved.

Name ______________________________ Class ______________ Date ____________

Assessment

Chapter Test

Biological Communities

In the space provided, write the letter of the term or phrase that best completes each statement or best answers each question.

______ **1.** Parasitism is considered a special case of
a. symbiosis.
b. predation.
c. competitive exclusion.
d. resource sharing.

______ **2.** Some interactions among species are the result of a long evolutionary history in which many of the participants
a. feed on one another.
b. compete with one another for resources.
c. evolve with one another over time.
d. cause other species to become extinct.

______ **3.** A long-term relationship in which both participating species benefit is known as
a. parasitism.
b. mutualism.
c. predation.
d. commensalism.

______ **4.** Thorns, spines, and secondary compounds are examples of
a. symbiosis.
b. coevolution.
c. plant/animal interaction.
d. plant defenses.

______ **5.** An example of an internal human parasite is a
a. hookworm.
b. flea.
c. mosquito.
d. tick.

______ **6.** Some ants and aphids are
a. parasites.
b. predators.
c. mutualistic.
d. commensalistic.

Copyright © by Holt, Rinehart and Winston. All rights reserved.

______ **7.** All the ways in which a jaguar interacts with its environment make up its
a. ecosystem.
b. niche.
c. habitat.
d. community.

______ **8.** Resources for which species compete include which of the following?
a. living space
b. light
c. food
d. All of the above

______ **9.** A niche is
a. an organism's functional role in an ecosystem.
b. the place where an organism lives.
c. the competition habits of an organism.
d. a major biological community.

______ **10.** The total niche an organism is potentially able to occupy within an ecosystem is its
a. realized niche.
b. habitat.
c. fundamental niche.
d. community.

______ **11.** When two protist species that use different resources were placed in the same culture tube, both of the species
a. became extinct.
b. became overpopulated.
c. competed for food resources.
d. were able to coexist.

______ **12.** The principle of competitive exclusion states that if two species are competing, the species that use the resource more efficiently will eventually
a. increase it use of the common resource.
b. grow exponentially and then level off.
c. eliminate the other species entirely.
d. eliminate the other species locally.

______ **13.** Areas that have a high degree of biodiversity
a. experience greater stability in adverse conditions.
b. demonstrate far more competition for resources among species.
c. generally have less living space per species.
d. experience an overall reduction in productivity.

Copyright © by Holt, Rinehart and Winston. All rights reserved.

______**14.** Predation can increase biodiversity by
- **a.** reducing the numbers of overcrowding prey species.
- **b.** increasing competition for resources among prey species.
- **c.** increasing the number of pollinators in an area.
- **d.** None of the above

______**15.** A biome's characteristics over time are NOT affected by
- **a.** average temperatures.
- **b.** localized disasters.
- **c.** amount of annual rainfall.
- **d.** degrees of latitude.

In the space provided, write the letter of the description that best matches the term or phrase.

______**16.** climate

______**17.** latitude

______**18.** temperature

______**19.** savanna

______**20.** desert

______**21.** tropical forest

______**22.** temperate forest

a. reduces the temperature and moisture conditions in an area as it increases

b. experiences very little differences in seasons

c. consists of two types—evergreen and deciduous

d. prevailing weather conditions in a given area

e. exist in tropical areas, but experience low rainfall

f. decreases as elevation increases

g. overriding characteristic is scarcity of water

Read each question, and write your answer in the space provided.

23. Describe the similarities and differences between the limnetic freshwater zone and the surface of the open sea.

____

____

____

____

____

____

Copyright © by Holt, Rinehart and Winston. All rights reserved.

24. How are freshwater habitats connected to terrestrial habitats?

____

____

25. Can two species occupy the same realized niche? Explain.

____

____

____

____

Copyright © by Holt, Rinehart and Winston. All rights reserved.

Name ______________________ Class ______________ Date ____________

Data Lab — DATASHEET FOR IN-TEXT LAB

Predicting How Predation Would Affect a Plant Species

Background

Grazing is the predation of plants by animals. Some plant species, such as *Gilia*, respond to grazing by growing new stems. Consider a field in which a large number of these plants are growing and being eaten by herbivores.

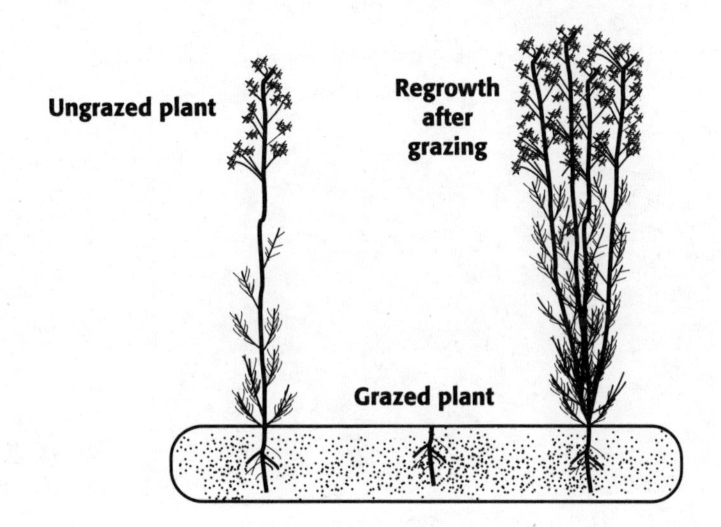

Analysis

1. **Identify** the plant that is likely to produce more seeds.

____

____

____

2. **Explain** how grazing affects this plant species.

____

____

____

3. **Evaluate** the significance to its environment of the plant's regrowth pattern.

____

____

____

Copyright © by Holt, Rinehart and Winston. All rights reserved.

4. **Hypothesize** how this plant species might be affected if individual plants did not produce new stems in response to grazing.

____

____

____

____

Copyright © by Holt, Rinehart and Winston. All rights reserved.

Name ________________________ Class ______________ Date ____________

Data Lab

DATASHEET FOR IN-TEXT LAB

Predicting Changes in a Realized Niche

Background

Two features of a niche that can be readily measured are the location where the species feeds and the size of its preferred prey. The darkest shade in the center of the graph indicates the prey size and feeding location most frequently selected by one bird species (called Species A).

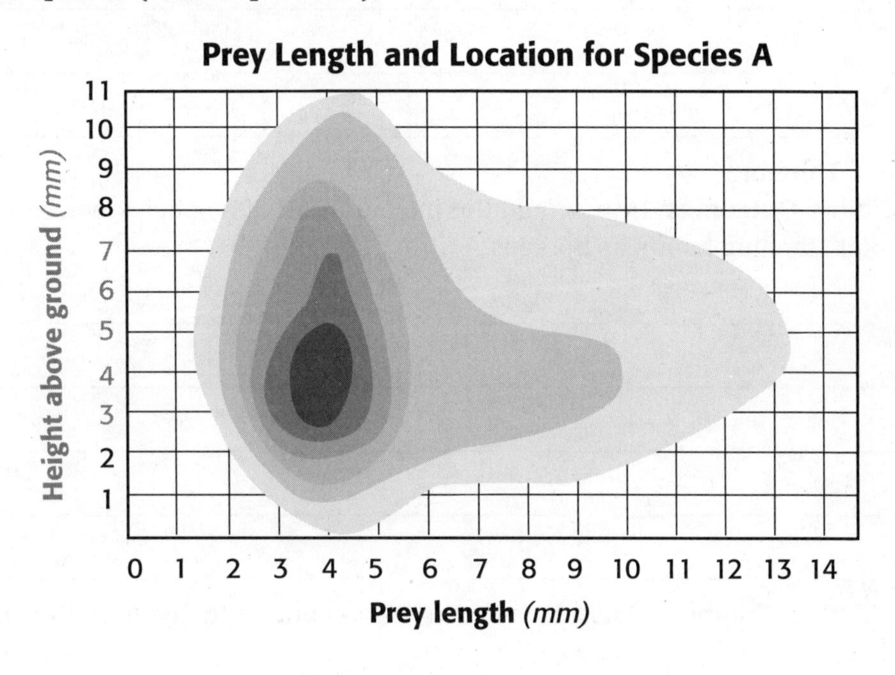

Analysis

1. **State** the range of lengths of Species A's preferred prey.

____

2. **Identify** the maximum height at which Species A feeds.

____

3. **Critical Thinking**
Predicting Outcomes Species B is introduced into Species A's feeding range. Species B has exactly the same feeding preferences, but it hunts at a slightly different time of day. How might this affect Species A?

____

____

____

____

Copyright © by Holt, Rinehart and Winston. All rights reserved.

4. **Interpreting Graphics** Species C is now introduced into Species A's feeding range. Species C feeds at the same time of day as Species A, but it prefers prey that are between 10 and 13 mm long. How might this affect Species A?

____

____

____

____

____

5. **Critical Thinking**
   Predicting Outcomes How would the introduction of a species with exactly the same feeding habits as Species A affect the graph?

____

____

____

____

____

6. **Interpreting Graphics** What does the lightest shade at the edge of the contour lines represent?

____

____

Copyright © by Holt, Rinehart and Winston. All rights reserved.

Name ______________________ Class ______________ Date __________

Quick Lab

DATASHEET FOR IN-TEXT LAB

Investigating Factors That Influence the Cooling of Earth's Surface

You can discover how the amount of water in an environment affects the rate at which that environment cools.

MATERIALS

- MBL or CBL system with appropriate software
- temperature probes
- test tubes
- beaker
- hot plate
- one-holed stoppers
- water
- sand
- test-tube tongs
- test-tube rack

Procedure

1. Set up an MBL/CBL system to collect and graph data from each temperature probe at 5-second intervals for 240 data points. Calibrate the probe using stored data.
2. Fill one test tube with water. Fill another test tube halfway with sand.
3. Place a temperature probe in the sand, and suspend another temperature probe at the same depth in the water, using one-holed stoppers to hold each temperature probe in place.
4. Place both test tubes in a beaker of hot water. Heat them to a temperature of about 70°C. **Caution: Hot water can burn skin.**
5. Using test-tube tongs, remove the test tubes and place them in the test-tube rack. Record the drop in temperature for 20 minutes.

Analysis

1. **Critical Thinking**
   Analyzing Results Did the two test tubes cool at the same rate? Offer an explanation for your observations.

   ____

   ____

2. **Critical Thinking**
   Predicting Outcomes In which biome—tropical rain forest or desert—would you expect the air temperature to drop most rapidly? Explain your answer.

   ____

   ____

Copyright © by Holt, Rinehart and Winston. All rights reserved.

Name ______________________ Class ______________ Date ____________

Skills Practice Lab

DATASHEET FOR IN-TEXT LAB

Observing How Brine Shrimp Select a Habitat

SKILLS

- Using scientific methods
- Collecting, organizing, and graphing data

OBJECTIVES

- **Observe** the behavior of brine shrimp.
- **Assess** the effect of environmental variables on habitat selection by brine shrimp.

MATERIALS

- clear, flexible plastic tubing
- metric ruler
- marking pen
- corks to fit tubing
- brine shrimp culture
- screw clamps
- test tubes with stoppers and test-tube rack
- pipet
- Petri dish
- Detain™ or methyl cellulose
- aluminum foil
- calculator
- fluorescent lamp or grow light
- funnel
- graduated cylinder or beaker
- hot-water bag
- ice bag
- pieces of screen
- tape

Before You Begin

Different organisms are adapted for life in different **habitats.** For example, **brine shrimp** are small crustaceans that live in salt lakes. Given a choice, organisms select habitats that provide the conditions (e.g., temperature, light, pH, salinity) to which they are adapted. In this lab, you will investigate habitat selection by brine shrimp and determine which environmental conditions they prefer.

1. Write a definition for each boldface term in the paragraph above.

____

____

____

2. Based on the objectives for this lab, write a question you would like to explore about habitat selection by brine shrimp.

____

____

Copyright © by Holt, Rinehart and Winston. All rights reserved.

Procedure

PART A: MAKING AND SAMPLING A TEST CHAMBER

1. Divide a piece of plastic tubing into 4 sections by making a mark at 12 cm, 22 cm, and 32 cm from one end. Label the sections *1*, *2*, *3*, and *4*.
2. Place a cork in one end of the tubing. Then transfer 50 mL of brine shrimp culture to the tubing. Place a cork in the open end of the tubing.
3. When you are ready to count shrimp, divide the tubing into four sections by placing a screw clamp at each mark on the tubing. *While someone holds the corks firmly in place*, first tighten the middle clamp and then the outer clamps.
4. Starting at one end, pour the contents of each section into a test tube labeled with the same number. After you empty a section, loosen the adjacent clamp and fill the next test tube.
5. Stopper one test tube, and invert it gently to distribute the shrimp. Use a pipet to transfer a 1 mL sample of shrimp culture to a Petri dish. Add a few drops of Detain™ to the sample. Count and record the number of live shrimp.
6. Repeat step 5 three more times for the same test tube. Record the average number of shrimp for this test tube.
7. Repeat steps 5 and 6 for each of the remaining test tubes.

PART B: DESIGN AN EXPERIMENT

8. Work with the members of your lab group to explore one of the questions written for step 2 of **Before You Begin.** To explore the question, design an experiment that uses the materials listed for this lab.

You Choose

As you design your experiment, decide the following:

- **a.** what question you will explore
- **b.** what hypothesis you will test
- **c.** how to set up your control
- **d.** how to expose the brine shrimp to the conditions you chose
- **e.** how long to expose the brine shrimp to the environmental conditions
- **f.** how you will set up your data table

Copyright © by Holt, Rinehart and Winston. All rights reserved.

9. Write a procedure for your group's experiment. Make a list of all the safety precautions you will take. Have your teacher approve your procedure and safety precautions before you begin the experiment.

10. Set up and conduct your group's experiment. Do *not* use water over 70°C, which can burn you. **CAUTION: If you are working with the hot-water bag, handle it carefully. If you are working with a lamp, do not touch the bulb. Light bulbs get very hot and can burn your skin.**

PART C: CLEANUP AND DISPOSAL

11. Dispose of broken glass in the designated waste container. Put brine shrimp in the designated container. Do not pour chemicals down the drain or put lab materials in the trash unless your teacher tells you to do so.

12. Clean up your work area and all lab equipment. Return lab equipment to its proper place. Wash your hands thoroughly before you leave the lab and after you finish all work.

Analyze and Conclude

1. Summarizing Results Make a bar graph of your data. Use graph paper to plot the environmental variable on the x-axis and the number of shrimp on the y-axis.

2. Analyzing Results How did the shrimp react to changes in the environment?

____

3. Analyzing Methods Why was a control necessary?

____

____

4. Analyzing Methods Why was it necessary to take many counts in each test tube (step 6 of Part A)?

____

____

5. Further Inquiry Write a new question about brine shrimp that could be explored with another investigation.

____

____

Copyright © by Holt, Rinehart and Winston. All rights reserved.

Exploration Lab

FIELD ACTIVITY

Life in a Pine Cone

Large numbers and many varieties of arthropods (insects, spiders, mites, centipedes, millipedes) can live in the structure of a single pine cone. The pine cones afford sanctuary for arthropod inhabitants and visitors to feed, seek mates, lay eggs, hide, or escape danger. Many of the arthropods on or inside cones probably gain access by moving from the soil, up the tree trunk, and out the branches to the cones. Others fly directly to the tree and cones. Entire life histories of some species are completed within cones before and after they fall to the ground and decompose.

The types and numbers of arthropods recovered from pine cones vary with pine species and weather conditions. Most species range in size from about 25 mm down to the size of the period at the end of this sentence.

To separate small arthropods from pine cones, a *Berlese funnel* can be used. This apparatus consists of a large funnel, the mouth of which may be covered with fine wire mesh. The funnel is set into a container. When material that contains small arthropods is placed in the funnel, the organisms collect in the container below while the rest of the material remains in the top of the funnel.

In this lab, you will collect pine cones from trees and from leaf litter and use a Berlese funnel to extract arthropods from the cones. You will identify and compare the arthropods from the two different pine cone microhabitats.

OBJECTIVES

Collect arthropods that live in two different pine cone microhabitats.

Identify the arthropods that live in the pine cone microhabitats.

Compare the diversity of arthropods living in the the two pine cone microhabitats.

MATERIALS

- beakers, small (2)
- Berlese funnel
- cloths, white, about 30 × 30 cm (2)
- felt tip marker
- field guide to arthropods
- filter paper
- forceps or fine paintbrush
- funnel
- gloves
- hand lens
- jar, large
- jars, small (2)
- lab apron
- lamp with 25 W bulb
- plastic bags (4)
- plastic hoop, cloth covered, for collecting cones
- rubbing alcohol
- safety goggles
- sieve, for collecting insects
- stereomicroscope
- trays (2)
- twist ties (4)

Copyright © by Holt, Rinehart and Winston. All rights reserved.

Procedure

PART 1: COLLECTING AND PRESERVING SPECIMENS

1. Collect at least five pine cones from leaf litter on the ground and five open pine cones from trees. Use a plastic hoop covered with cloth to dislodge the cones from trees. Place the leaf-litter pine cones and the tree pine cones in separate plastic bags. Use a meter to label each bag with the location from which the cones were taken.
2. Use an insect sieve to collect small insects and other small arthropods from the cones. As shown in **Figure 1,** the sieve has a rigid wire ring top, cloth sides, and a wire mesh bottom. Place the sieve over a white cloth, place collected pine cones from one of your bags on the wire mesh, and shake.
3. When you finish shaking the cones, place the white cloth containing the materials you have dislodged in a plastic bag, and tie it securely. Mark the bag with the location (leaf litter or trees) from which the materials were obtained.
4. Use forceps or a fine paintbrush to collect tiny arthropods remaining inside the cones. Carefully pick out or brush the animals and place them in the plastic bag with the other materials you collected in step 3. Tie the bag securely.

FIGURE 1 INSECT SIEVE

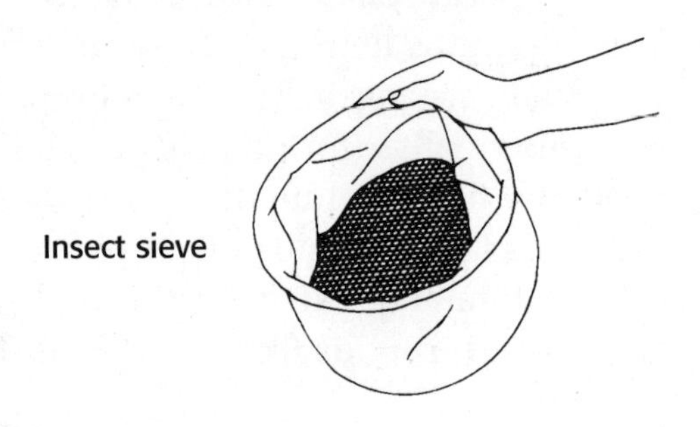

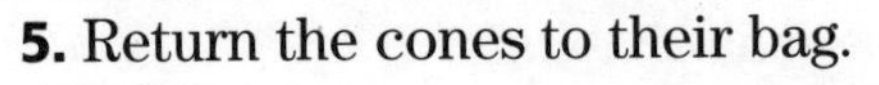

5. Return the cones to their bag.
6. Repeat steps 2-5, using the cones from the other location.
7. Carry your bags of cones and bags of collected materials back to the lab.
8. In the lab, place a small jar inside a large jar, as shown in **Figure 2.** Place the jars under a Berlese funnel. Then place the funnel under a lamp with a 25 W light bulb.
9. Empty one of your bags of cones into the funnel. Turn on the lamp. **CAUTION: Be sure the lamp is at least 5 cm from the cones.** The heat from the bulb will dry the material in the funnel from the top down. As the material warms and dries, the organisms present will move to lower levels and fall through the neck of the funnel into the small jar.

FIGURE 2 APPARATUS FOR DRYING MATERIALS

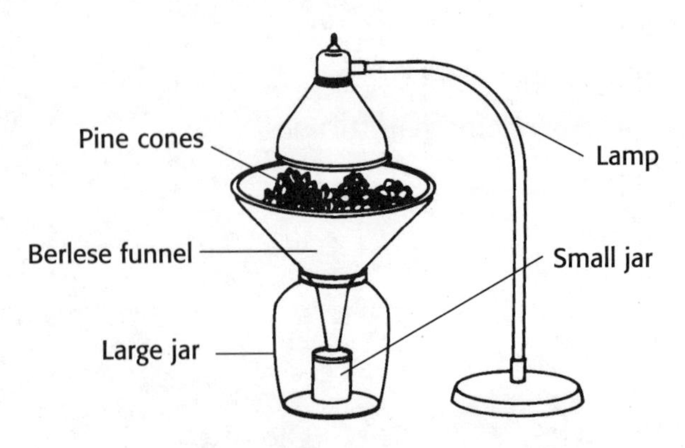

Copyright © by Holt, Rinehart and Winston. All rights reserved.

10. Put on safety goggles, gloves, and a lab apron.

11. Fill a small beaker $\frac{1}{4}$ full with rubbing alcohol. **CAUTION: Rubbing alcohol is poisonous if swallowed. Avoid contact with eyes.** Transfer the material you collected in the small jar in step 9 to the beaker with rubbing alcohol. The alcohol will preserve the organisms.

12. Add to the alcohol all the other materials you collected (in steps 3 and 4) from the same cones.

13. Separate the materials from the rubbing alcohol by pouring the mixture into a funnel lined with filter paper held over a small jar. Gently transfer the preserved organisms remaining on the filter paper onto a tray. Label the tray with the location (leaf litter or trees) from which the organisms were obtained.

14. Repeat steps 9-13, using your other bag of cones.

PART 2: EXAMINING AND IDENTIFYING SPECIMENS

15. Each of the locations from which you collected cones is a separate microhabitat for arthropods. Examine under a stereomicroscope the preserved arthropods from one of the microhabitats.

16. Group together organisms that are alike, and record them as taxa, or discrete groups of organisms. You do not have to know what each organism is, only that each group has identical individuals.

17. Count the number of individuals in each taxon, and record your tally in the appropriate column in **Table 1.**

TABLE 1 ARTHROPODS COLLECTED FROM TWO MICROHABITATS

Habitat 1 Cones in leaf litter			Habitat 2 Cones on trees		
Taxon	Number of individuals	Name	Taxon	Number of individuals	Name
1			1		
2			2		
3			3		
4			4		
5			5		

18. Repeat steps 15-17, using the arthropods from the other microhabitat.

19. Use the dichotomous key in **Figure 3** to identify the arthropods you grouped in **Table 1.** You may also want to consult a field guide. In **Table 1,** write the names of the arthropods you are able to identify.

Copyright © by Holt, Rinehart and Winston. All rights reserved.

20. Dispose of your materials according to your teacher's instructions.

21. Clean up your work area, and wash your hands before leaving the lab.

FIGURE 3 DICHOTOMOUS KEY FOR IDENTIFYING ARTHROPODS

1.	a.	Six legs	Go to 2
	b.	Eight or more legs	Go to 3
2.	a.	No wings	Go to 9
	b.	Wings	Flying insect (Insecta)
3.	a.	Eight legs	Go to 4
	b.	More than eight legs	Go to 7
4.	a.	No cephalothorax	Go to 5
	b.	Cephalothorax present	Go to 6
5.	a.	Long legs	Harvestman (Arachnida)
	b.	Short legs	Mite (Arachnida)
6.	a.	Pincers	Pseudoscorpion (Arachnida)
	b.	No pincers	Spider (Arachnida)
7.	a.	Oval body	Pill bug (Crustacea)
	b.	Long, thin body	Go to 8
8.	a.	One pair of legs per segment	Centipede (Chilipoda)
	b.	Two pairs of legs per segment	Millipede (Diploda)
9.	a.	Chewing mouthparts	Go to 10
	b.	Sucking mouthparts	Go to Scales, Aphid, Leaf hopper, or Cicada (Insecta)
10.	a.	No appendages on abdomen	Go to 11
	b.	Appendages on abdomen	Go to 14
11.	a.	Narrow connection between thorax and abdomen	Ant
	b.	Thick connection between thorax and abdomen	Go to 12
12.	a.	Short antenna	Termite (Insecta)
	b.	Long antenna	Go to 13
13.	a.	All legs the same length	Plant lice (Insecta)
	b.	Longer third pair of legs	Grasshopper, cricket (Insecta)
14.	a.	Hairlike tail	Silverfish (Insecta)
	b.	Forked tail	Springtail (Insecta)

Copyright © by Holt, Rinehart and Winston. All rights reserved.

Analysis

1. **Organizing Data** The *diversity index* of a habitat is a number derived from the following formula. In the space provided, use this formula to calculate the diversity index of each microhabitat.

$$\text{Diversity} = \frac{N\,(N - 1)}{\Sigma n\,(n \quad 1)}$$

Where: N = total number of individuals of all species
n = number of individuals of a species
Σ = the sum of all individuals sampled, as $n\,(n - 1)$

Diversity = the number of species in a defined area

2. **Analyzing Data** Compare the diversity of arthropods in the two different pine cone microhabitats.

____

____

____

____

Copyright © by Holt, Rinehart and Winston. All rights reserved.

3. Classifying Which arthropods did you identify? To which classes do they belong?

____

____

____

____

____

Conclusions

1. Drawing Conclusions How would you account for the difference in diversity of arthropods in the two microhabitats?

____

____

____

Extensions

1. Research and Communications Preserve specimens collected from pine cones permanently for later identification by putting them into a jar with a 50–70% alcohol solution. Find out how to ship the specimens to a systematist for identification. If you ship them, be sure to keep one of each species for future reference.

2. Designing Experiments Design a study of organisms in another type of microhabitat that can be carried out in an environment near you. For example, you might collect small arthropods from trees other than pines. You might study the variety of organisms in an open field or along the shore of a lake or stream.

Copyright © by Holt, Rinehart and Winston. All rights reserved.

Name ______________________ Class ______________ Date ____________

Skills Practice Lab

RESEARCH

Examining Owl Pellets

Owls are *raptors*, or birds of prey. They catch their prey, including small birds and rodents, and swallow them whole, since owls do not have teeth for grinding. Enzymatic juices in the owl's digestive system break down the body tissues of the prey but leave the bony materials, hair, and feathers undigested. Depending on the prey eaten, the undigested portions may include beaks, claws, scales, or insect exoskeletons. These materials have little nutritional value and must be eliminated from the owl's body. They form a *lump* called a *pellet*. Pellets begin forming within the digestive tract of the owl as soon as the prey is swallowed. The pellets are then coughed up, or regurgitated, and the owl begins feeding once again.

Scientists take advantage of this adaptation by collecting owl pellets and analyzing their contents. Owl pellets are dried and either fumigated (treated with chemicals) or sterilized so that their contents can be examined safely. Since owls are not selective feeders, the pellets can be used to estimate the diversity of available prey. The contents are also a direct indicator of the animals on which an owl has fed, which is information that is crucial for species management and protection.

In this lab, you will dissect pellets of barn owls from two different habitats—the Northwest and the Southeast United States. Based on your findings, you will compare the diets of these barn owls.

OBJECTIVES

- **Identify** the species of prey eaten by the Northwestern and Southeastern barn owls from the remains found in the owls' pellets.
- **Compute** the total biomass of each prey and the cumulative total biomass, using class data for each type of barn-owl pellet.
- **Compare** the diets of the Northwestern and Southeastern barn owls, using the data.

MATERIALS

- dissecting needle
- dissecting tray
- forceps
- gloves
- lab apron
- metric ruler
- paper, white
- pellet, Northwestern barn owl
- pellet, Southeastern barn owl
- safety goggles
- tape or glue

Copyright © by Holt, Rinehart and Winston. All rights reserved.

Procedure

PART 1: DISSECTING AN OWL PELLET

1. Put on safety goggles, gloves, and a lab apron. **CAUTION: Do not touch your hands to your face or mouth during this lab.**
2. Place an owl pellet in a dissecting tray. Remove the pellet from the aluminum foil casing.
3. Use a dissecting needle and forceps to carefully break apart the owl pellet. **CAUTION: Use sharp instruments with extreme care.** Remove the fur and feathers from the bones. *Note: Be careful to avoid damaging small bones while you are pulling the pellet apart.*
4. As the bones are uncovered, use forceps to carefully place them on a sheet of paper. Take care to remove all skulls and bones from the fur mass. You will identify the animals the owl has eaten mainly by the skulls, mandibles (jaws), and teeth.

   Sometimes undigested beetles and pill bugs are found in owl pellets. These small animals find the expelled pellets and use them as a food source and nursery for their eggs and larvae. Therefore, these organisms should not be included as owl prey.
5. Use the diagrams of the bird and mammal skeletons shown in **Figures 1** and **2,** to help you distinguish among different types of bones.

FIGURE 1 GENERALIZED BIRD SKELETON

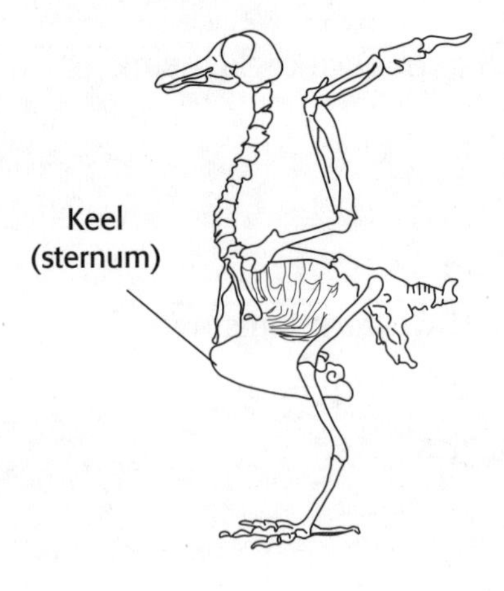

FIGURE 2 GENERALIZED MAMMAL SKELETON

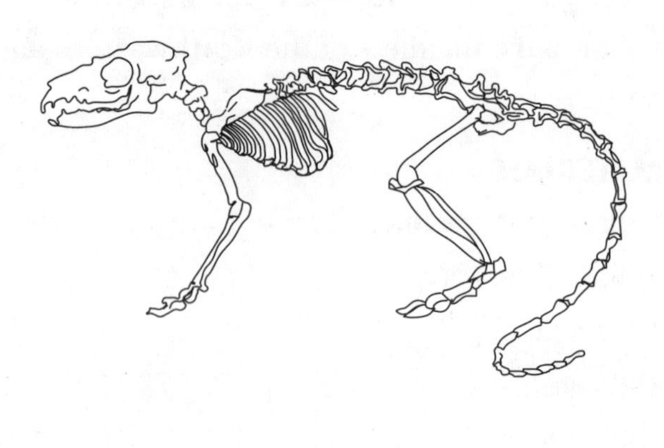

Copyright © by Holt, Rinehart and Winston. All rights reserved.

6. Assemble similar bone parts to see how many prey types are represented in the pellet. Count the number of skulls to determine how many prey were in the pellet. Refer to **Figure 3** to help identify types of skulls.

FIGURE 3 SKULL COMPARISONS

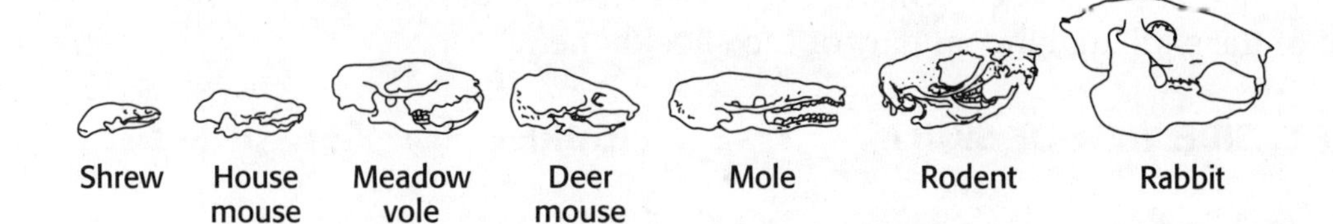

7. Reassemble the skeletons by laying them out on a sheet of paper with appendages arranged and spread out to the side. Glue or tape the assembled skeletons to the paper.

- How many skeletons were you able to assemble from your owl pellet?

____

____

8. Repeat steps 2–7 for the second owl pellet.

PART 2: USING A DICHOTOMOUS KEY TO IDENTIFY PREY

The contents of an owl pellet can be identified with a *dichotomous key*, a tool used to identify an object or an organism. A dichotomous key has a series of statements or questions that compare contrasting characteristics among a group of items or organisms. For example, suppose you want to identify a common U.S. coin by using a dichotomous key. The key might be similar to the one shown in **Figure 4.** Compare the first pair of statements and determine which one best fits the coin you are trying to identify. After you pick one of the paired statements, you will be directed to another paired statement until you reach an answer.

FIGURE 4 DICHOTOMOUS KEY OF COINS

1.	Coin edge smooth	Go to 2
	Coin edge grooved	Go to 3
2.	Coin copper in color	penny
	Coin silver in color	nickel
3.	Picture of Roosevelt on front	dime
	Picture of Washington on front	quarter

Copyright © by Holt, Rinehart and Winston. All rights reserved.

9. Use **Figures 5** and **6,** showing the side and top views of a skull, along with the following information to become familiar with terminology used in a dichotomous key of small mammal skulls.

A gap between the incisors (front teeth) and cheek teeth (molars and premolars) is called a *diastema.* The *infraorbital canal* refers to an opening just below the eye socket. The *zygomatic arch* is a ridge of bone located on the side of the skull making up part of the cheekbone.

FIGURE 5 SIDE VIEW OF SKULL

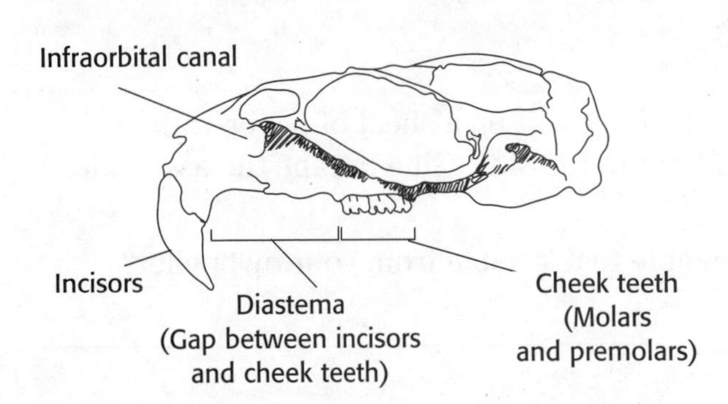

FIGURE 6 TOP VIEW OF SKULL

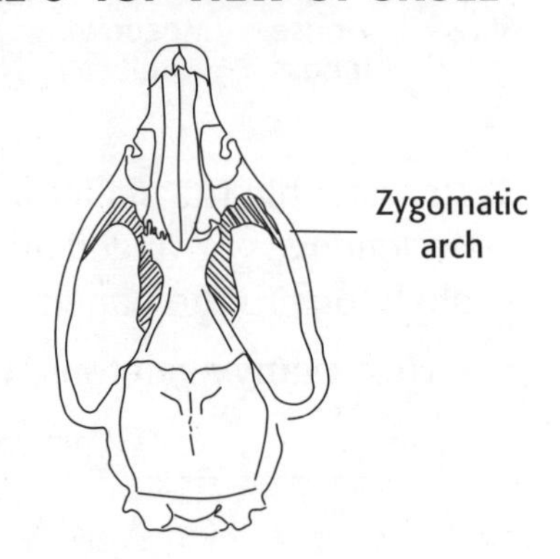

10. Use **Figures 7** and **8** and the following information to help you determine the length of the mandible, or lower jaw.

The bottom view of the skull is actually a view of the roof of the mouth. When determining the posterior edge of the *palate,* look for the position of the *posterior palatine foramina.* The palate ends beyond the posterior palatine foramina.

The skull on the left in **Figure 8** shows the posterior edge of the palate ending even with the last cheek teeth. The right-hand skull shows the palate ending beyond the last cheek teeth.

FIGURE 7 SIDE VIEW OF MANDIBLE (JAW)

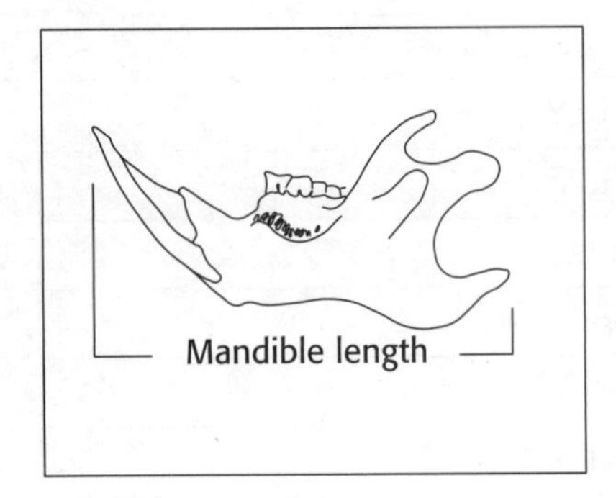

FIGURE 8 BOTTOM VIEW OF SKULLS

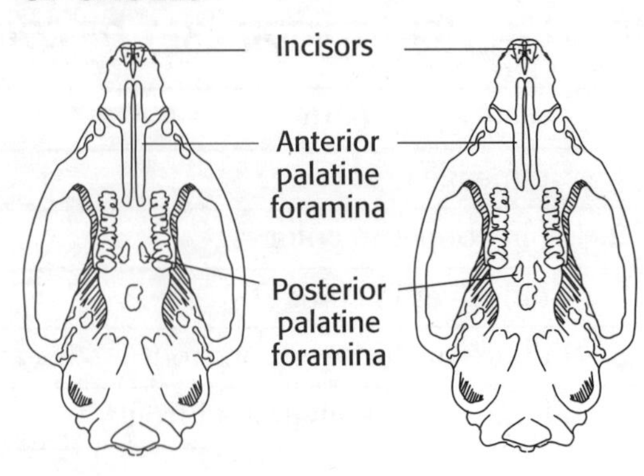

Copyright © by Holt, Rinehart and Winston. All rights reserved.

11. To determine cheek teeth types or incisor types, refer to **Figures 9** and **10.**

FIGURE 9 CHEEK TEETH TYPES

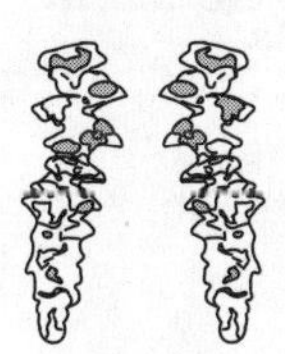
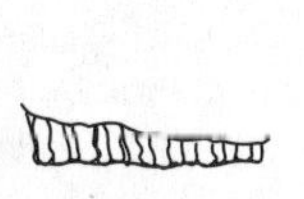
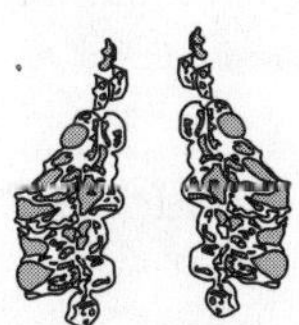
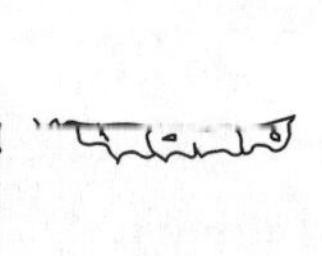
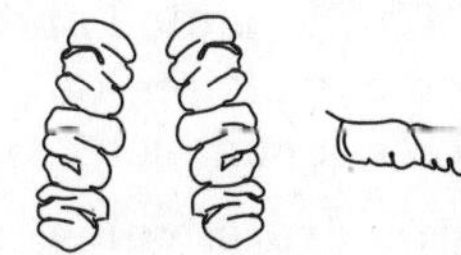

Acute (bottom view)	Acute (side view)	Unicusp (bottom view)	Unicusp (side view)	Lobed s-shaped (bottom view)	Lobed s-shaped (side view)

FIGURE 10 INCISOR TYPES

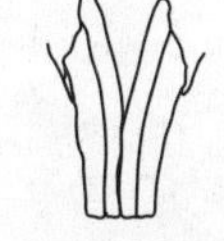
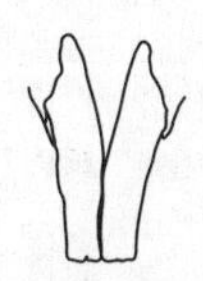

Grooved incisor (front view)	Smooth incisor (front view)	Notched incisor (front view)	Unnotched incisor (front view)

12. Use the dichotomous key in **Figure 11** and the diagrams in **Figures 5-10** to identify the skulls of small mammals found in your owl pellets. In **Table 1** in the *Number found in pellet* column, record the number of each type of mammal skull found in the Northwestern barn-owl pellet and in the Southeastern barn-owl pellet.

13. Any other animal remains you find in your pellets will be from a bird, a bat, or an insect. List them as "other prey" in **Table 1.**

- How many different species of animals did you find in each owl pellet?

____

14. Your teacher will gather class totals for each type of prey in both types of owl pellets. Record these in the *Total number in all pellets* column in **Table 1.**

Copyright © by Holt, Rinehart and Winston. All rights reserved.

FIGURE 11 DICHOTOMOUS KEY OF SMALL MAMMAL SKULLS

No gap (diastema) between incisors and cheek teeth ....Order: Insectivora
Gap (diastema) between incisors and cheek teeth ....Order: Rodentia

Order Insectivora (moles and shrews)

Zygomatic arch complete; skull flat and broad ....*Scapanus* (mole)
Zygomatic arch not complete; skull not flat and broad ....*Sorex* (shrew)

Order Rodentia (rats, voles, and mice)

1. Infraorbital canal not present ....Go to 2
Infraorbital canal present ....Go to 3

2. Upper incisors distinctly grooved ....*Perognathus* (mouse)
Upper incisors not distinctly grooved ....*Thomomys* (pocket gopher)

3. Skull flat and broad; cheek teeth acutely angled and may appear as one continuous tooth ....*Microtus* (vole)
Skull generally rounded; cheek teeth lobed or rounded and easily distinguished individually ....Go to 4

4. Upper incisors distinctly grooved..........*Reithrodontomys* (harvest mouse)
Upper incisors not distinctly grooved ....Go to 5

5. Posterior edge of palate ending even with or last cheek teeth; cheek teeth not capped with enamel ....*Peromyscus* (deer mouse)
Posterior edge of palate ending beyond last cheek teeth; cheek teeth capped with enamel..........Go to 6

6. Upper incisors notched, mandible length less than 16 mm ....*Mus* (house mouse)
Upper incisors not notched, mandible length greater than 18 mm..........*Rattus* (rat)

15. To compute the total biomass of a prey, multiply the number in the *Total number in all pellets* column by the number listed in the *Prey biomass* column. For example:

If five pocket gophers *(Thomomys)* were recorded as the *Total number in all pellets*, and from the chart the prey biomass is 150 g, then

$$5 \times 150 \text{ g} = 750 \text{ g} = \text{total biomass}$$

Compute totals in this way for all species found, and enter your answers in the *Total biomass* column. Calculate *Cumulative total biomass* by summing all data in the *Total biomass* column.

16. Dispose of your materials according to your teacher's instructions.

17. Clean up your work area, and wash your hands before leaving the lab.

Copyright © by Holt, Rinehart and Winston. All rights reserved.

Name ______________________ Class ______________ Date ____________

Examining Owl Pellets *continued*

TABLE 1 ANALYSIS OF BARN-OWL PREY

Prey	Prey biomass (in grams)	Number found in pellet		Total number in all pellets		Total biomass (in grams)	
		N	S	N	S	N	S
Pocket gopher *Thomomys*	150						
Rat *Rattus*	150						
Vole *Microtus*	40						
Mice *Peromyscus* *Mus* *Reithrodontomys* *Perognathus*	 22 18 12 25						
Mole *Scapanus*	55						
Shrew *Sorex*	4						
Other prey bats birds insects	 7 15 1						
Cumulative total biomass (in grams)							

Key: N = Northwestern barn owl; S = Southeastern barn owl

Analysis

1. **Describing Events** How do your data compare with the class's pooled data?

____

____

2. **Analyzing Data** If a barn owl needs 120 g of food per day, how many *Sorex* would it need to capture? How many *Microtus?*

____

____

Copyright © by Holt, Rinehart and Winston. All rights reserved.

3. Analyzing Data Compare the diets of the Northwestern and Southeastern barn owls by comparing the total biomass of each prey to the cumulative total biomass for each type of barn owl.

Conclusions

1. Drawing Conclusions In what way might the formation of owl pellets increase an owl's chances of survival in an ecosystem?

2. Making Predictions How would a sudden decrease in the *Sorex* population affect a barn-owl population? What effect would a decrease in the *Microtus* population have?

3. Drawing Conclusions Assume an owl eats one hundred 1-gram insects and one 100-gram rat. Did the insects or rat contribute more to the owl's diet? Hint: How does foraging time affect this outcome?

Extension

Research and Communications Predators such as owls are subjected to poisons that have been concentrated through the food chain. Some species of owls are considered endangered, threatened by pesticides and heavy metals that contaminate the foods eaten by their prey. Research this topic and explain how a grain crop contaminated with a compound containing lead can affect an owl population.

Copyright © by Holt, Rinehart and Winston. All rights reserved.

Name ______________________ Class ______________ Date ____________

Data Lab

DATASHEET FOR IN-TEXT LAB

Predicting How Predation Would Affect a Plant Species

Background

Grazing is the predation of plants by animals. Some plant species, such as *Gilia*, respond to grazing by growing new stems. Consider a field in which a large number of these plants are growing and being eaten by herbivores.

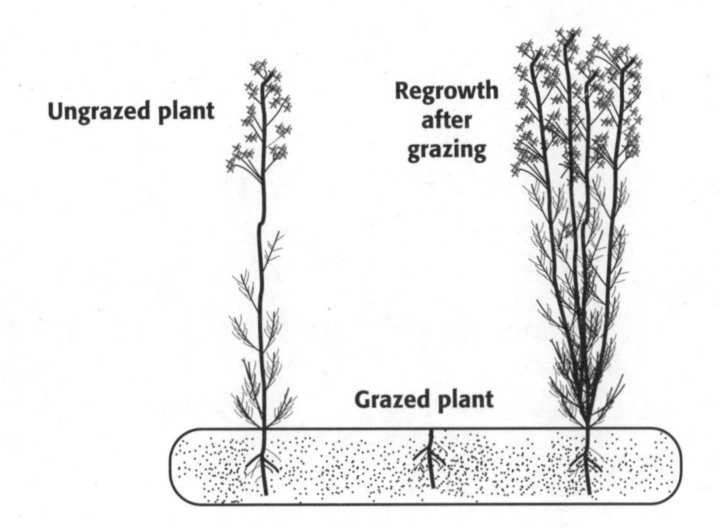

Analysis

1. **Identify** the plant that is likely to produce more seeds.

   The grazed plant would most likely produce more seeds because it would have more stems and flowers.

2. **Explain** how grazing affects this plant species.

   Because grazing leads to dense regrowth and the production of more flower heads, the grazed plant would produce more offspring.

3. **Evaluate** the significance to its environment of the plant's regrowth pattern.

   Dense regrowth and the production of more flower heads may allow this plant to spread in its environment and outcompete other plants.

Copyright © by Holt, Rinehart and Winston. All rights reserved.

Name ______________________ Class ______________ Date ____________

Predicting How Predation Would Affect a Plant Species *continued*

4. **Hypothesize** how this plant species might be affected if individual plants did not produce new stems in response to grazing.

If new stems were not produced in response to grazing, the grazed plants would produce few, if any, seeds. Over the years, the plant might become rare or extinct in areas of heavy grazing.

Copyright © by Holt, Rinehart and Winston. All rights reserved.

Name ______________________ Class ______________ Date ______________

Data Lab

DATASHEET FOR IN-TEXT LAB

Predicting Changes in a Realized Niche

Background

Two features of a niche that can be readily measured are the location where the species feeds and the size of its preferred prey. The darkest shade in the center of the graph indicates the prey size and feeding location most frequently selected by one bird species (called Species A).

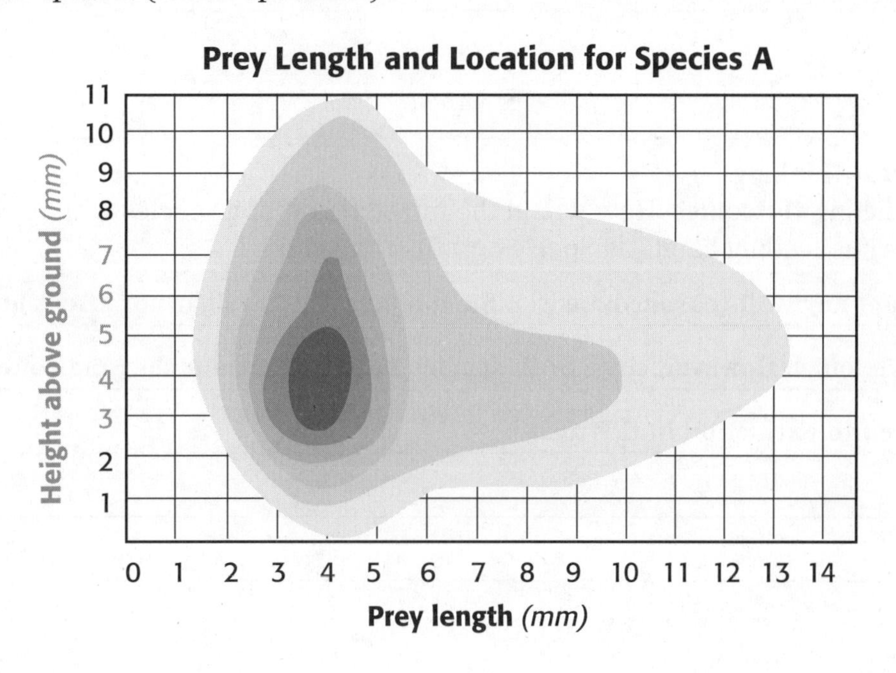

Analysis

1. **State** the range of lengths of Species A's preferred prey.

   Most selected prey are approximately 3.5 to 4.5 mm.

2. **Identify** the maximum height at which Species A feeds.

   Maximum feeding height is nearly 11 m.

3. **Critical Thinking**
   Predicting Outcomes Species B is introduced into Species A's feeding range. Species B has exactly the same feeding preferences, but it hunts at a slightly different time of day. How might this affect Species A?

   Even though it is feeding at a different time of day, Species B might reduce the prey available to Species A, since it has the same feeding preference.

Copyright © by Holt, Rinehart and Winston. All rights reserved.

Name ______________________ Class ______________ Date ____________

Predicting Changes in a Realized Niche *continued*

4. **Interpreting Graphics** Species C is now introduced into Species A's feeding range. Species C feeds at the same time of day as Species A, but it prefers prey that are between 10 and 13 mm long. How might this affect Species A?

   Species C would reduce Species A's realized niche by competing with Species A for large prey. Since Species A prefers smaller species, however, competition from Species C would be minimal.

5. **Critical Thinking**
   Predicting Outcomes How would the introduction of a species with exactly the same feeding habits as Species A affect the graph?

   Accept any well-reasoned answer. Sample answer: It might not affect how the graph looks. However, since both species have the same niche, one could be forced to extinction in this area.

6. **Interpreting Graphics** What does the lightest shade at the edge of the contour lines represent?

   The lightest shade represents the combination of feeding height and prey length least frequently selected but still exploited by Species A.

Copyright © by Holt, Rinehart and Winston. All rights reserved.

Name ____________________ Class ______________ Date ____________

Quick Lab

DATASHEET FOR IN-TEXT LAB

Investigating Factors That Influence the Cooling of Earth's Surface

You can discover how the amount of water in an environment affects the rate at which that environment cools.

MATERIALS

- MBL or CBL system with appropriate software
- temperature probes
- test tubes
- beaker
- hot plate
- one-holed stoppers
- water
- sand
- test-tube tongs
- test-tube rack

Procedure

1. Set up an MBL/CBL system to collect and graph data from each temperature probe at 5-second intervals for 240 data points. Calibrate the probe using stored data.
2. Fill one test tube with water. Fill another test tube halfway with sand.
3. Place a temperature probe in the sand, and suspend another temperature probe at the same depth in the water, using one-holed stoppers to hold each temperature probe in place.
4. Place both test tubes in a beaker of hot water. Heat them to a temperature of about 70°C. **Caution: Hot water can burn skin.**
5. Using test-tube tongs, remove the test tubes and place them in the test-tube rack. Record the drop in temperature for 20 minutes.

Analysis

1. **Critical Thinking**
   Analyzing Results Did the two test tubes cool at the same rate? Offer an explanation for your observations.

   The sand should cool faster, as water has a greater capacity to store heat and therefore a slower cooling rate.

2. **Critical Thinking**
   Predicting Outcomes In which biome—tropical rain forest or desert—would you expect the air temperature to drop most rapidly? Explain your answer.

   The desert temperature would drop more rapidly, due to the lack of water in the ground and the atmosphere.

Copyright © by Holt, Rinehart and Winston. All rights reserved.

Name ______________________ Class ______________ Date ______________

Skills Practice Lab

DATASHEET FOR IN-TEXT LAB

Observing How Brine Shrimp Select a Habitat

SKILLS

- Using scientific methods
- Collecting, organizing, and graphing data

OBJECTIVES

- **Observe** the behavior of brine shrimp.
- **Assess** the effect of environmental variables on habitat selection by brine shrimp.

MATERIALS

- clear, flexible plastic tubing
- metric ruler
- marking pen
- corks to fit tubing
- brine shrimp culture
- screw clamps
- test tubes with stoppers and test-tube rack
- pipet
- Petri dish
- Detain™ or methyl cellulose
- aluminum foil
- calculator
- fluorescent lamp or grow light
- funnel
- graduated cylinder or beaker
- hot-water bag
- ice bag
- pieces of screen
- tape

Before You Begin

Different organisms are adapted for life in different **habitats.** For example, **brine shrimp** are small crustaceans that live in salt lakes. Given a choice, organisms select habitats that provide the conditions (e.g., temperature, light, pH, salinity) to which they are adapted. In this lab, you will investigate habitat selection by brine shrimp and determine which environmental conditions they prefer.

1. Write a definition for each boldface term in the paragraph above.

   habitats–areas where organisms live and to which they are adapted

   brine shrimp–small crustaceans that live in salt lakes

2. Based on the objectives for this lab, write a question you would like to explore about habitat selection by brine shrimp.

   Answers will vary. For example: Will brine shrimp prefer a warm habitat or a cool habitat?

Copyright © by Holt, Rinehart and Winston. All rights reserved.

Name ______________________ Class ______________ Date ______________

Procedure

PART A: MAKING AND SAMPLING A TEST CHAMBER

1. Divide a piece of plastic tubing into 4 sections by making a mark at 12 cm, 22 cm, and 32 cm from one end. Label the sections *1*, *2*, *3*, and *4*.
2. Place a cork in one end of the tubing. Then transfer 50 mL of brine shrimp culture to the tubing. Place a cork in the open end of the tubing.
3. When you are ready to count shrimp, divide the tubing into four sections by placing a screw clamp at each mark on the tubing. *While someone holds the corks firmly in place*, first tighten the middle clamp and then the outer clamps.
4. Starting at one end, pour the contents of each section into a test tube labeled with the same number. After you empty a section, loosen the adjacent clamp and fill the next test tube.
5. Stopper one test tube, and invert it gently to distribute the shrimp. Use a pipet to transfer a 1 mL sample of shrimp culture to a Petri dish. Add a few drops of Detain™ to the sample. Count and record the number of live shrimp.
6. Repeat step 5 three more times for the same test tube. Record the average number of shrimp for this test tube.
7. Repeat steps 5 and 6 for each of the remaining test tubes.

PART B: DESIGN AN EXPERIMENT

8. Work with the members of your lab group to explore one of the questions written for step 2 of **Before You Begin.** To explore the question, design an experiment that uses the materials listed for this lab.

A sample procedure and data table appear in the TE for this lab.

You Choose
As you design your experiment, decide the following:

- **a.** what question you will explore
- **b.** what hypothesis you will test
- **c.** how to set up your control
- **d.** how to expose the brine shrimp to the conditions you chose
- **e.** how long to expose the brine shrimp to the environmental conditions
- **f.** how you will set up your data table

Copyright © by Holt, Rinehart and Winston. All rights reserved.

Name ______________________ Class ______________ Date ______________

Observing How Brine Shrimp Select a Habitat *continued*

9. Write a procedure for your group's experiment. Make a list of all the safety precautions you will take. Have your teacher approve your procedure and safety precautions before you begin the experiment.
10. Set up and conduct your group's experiment. Do *not* use water over 70°C, which can burn you. **CAUTION: If you are working with the hot-water bag, handle it carefully. If you are working with a lamp, do not touch the bulb. Light bulbs get very hot and can burn your skin.**

PART C: CLEANUP AND DISPOSAL

11. Dispose of broken glass in the designated waste container. Put brine shrimp in the designated container. Do not pour chemicals down the drain or put lab materials in the trash unless your teacher tells you to do so.
12. Clean up your work area and all lab equipment. Return lab equipment to its proper place. Wash your hands thoroughly before you leave the lab and after you finish all work.

Analyze and Conclude

1. **Summarizing Results** Make a bar graph of your data. Use graph paper to plot the environmental variable on the x-axis and the number of shrimp on the y-axis.
2. **Analyzing Results** How did the shrimp react to changes in the environment?

   Answers will vary depending on the species of *Artemia* used.
3. **Analyzing Methods** Why was a control necessary?

   The control was necessary to show that the brine shrimp did not inherently prefer one part of the tube.
4. **Analyzing Methods** Why was it necessary to take many counts in each test tube (step 6 of Part A)?

   Many counts were taken to make allowances for variations in populations and to provide data for calculating an average.
5. **Further Inquiry** Write a new question about brine shrimp that could be explored with another investigation.

   Answers will vary. For example: How do brine shrimp react to water movement?

Copyright © by Holt, Rinehart and Winston. All rights reserved.

Exploration Lab

FIELD ACTIVITY

Life in a Pine Cone

Teacher Notes

TIME REQUIRED Two 45-minute periods

SKILLS ACQUIRED

Collecting data
Identifying patterns
Interpreting
Organizing and analyzing data

RATINGS Easy ← 1 2 3 4 → Hard

Teacher Prep–2
Student Setup–3
Concept Level–2
Cleanup–2

THE SCIENTIFIC METHOD

Make Observations Part 2 of the Procedure requires students to make observations.

Analyze the Results Analysis question 2 requires students to analyze their results.

Draw Conclusions Conclusions question 1 asks students to draw conclusions from their data.

MATERIALS

Materials for this lab can be purchased from WARD'S. See the *Master Materials List* for ordering instructions.

A suction-tube aspirator can be used instead of a fine paintbrush or forceps to pick out tiny arthropods from the cones. A suitable aspirator is available from WARD'S.

A Berlese funnel consists of a wide mouth funnel with a fine mesh screen at the bottom. Students can make their own Berlese funnel by using a large plastic (kitchen-type) funnel or the inverted top half of a plastic jug. A piece of fine mesh screen should be stapled or tacked over the inside of the funnel neck to keep litter from falling through.

Pines from many different regions of the United States can be used for this lab. These include loblolly pine *(Pinus taeda)* and shortleaf pine *(P. echinata)* from East Texas, ponderosa pine *(P. ponderosa)* from the Western states, red pine *(P. resinosa)* from the Great Lakes region, eastern white pine *(P. strobus)* from the Northeast, Virginia pine *(P. Virginiana)* in the Piedmont, and Choctawhatchee sand pine (*P. clausa*, var. choctawhatchee) from the Southeast.

Copyright © by Holt, Rinehart and Winston. All rights reserved.

SAFETY CAUTIONS

- Make sure that students are properly dressed for a collecting trip in the field.
- Assess the area for hazards, and review safety procedures with students. Caution students not to touch animals or poisonous plants.
- Rubbing alcohol (70% isopropyl alcohol) is a flammable liquid; avoid open flames, excessive heat, sparks, and other potential ignition sources. Rubbing alcohol is poisonous by ingestion and is an eye and skin irritant. In case of contact, flush affected areas with water for 15 minutes, including under eyelids; rinse mouth with water. Get prompt medical attention; induce vomiting (only in a conscious person).
- Pine cones and leaf litter are fire hazards. Closely monitor the heating of these materials with the light bulb.

DISPOSAL

- Cones that are in the same condition as they were when collected can be returned to the natural environment. Dispose of all other cones and the preserved organisms in a biohazard bag.
- Dilute small volumes (less than 250 mL) of rubbing alcohol in a ratio of 1 part solution to 20 parts water in a beaker. Place the beaker of diluted solution in a sink, and run water to overflowing for 10 minutes, flushing to a sanitary sewer.

TECHNIQUES TO DEMONSTRATE

Show students how to use the dichotomous key provided to identify their specimens.

TIPS AND TRICKS

Preparation

This lab works best in groups of two to four students.

Find out if you will be allowed to remove pine cones from the area where you wish to collect. Some areas are preserves and do not permit removing pine cones.

Discuss with students the science of taxonomy and the use of classification keys.

Procedure

If pine cones found on the ground are not embedded in leaf litter, an alternative microhabitat may be chosen.

Part 2 of the lab can be conducted several days later than Part 1 if necessary.

Be sure students understand the formula used to calculate diversity. Sample calculations are provided from the sample data given.

Copyright © by Holt, Rinehart and Winston. All rights reserved.

Name ______________________ Class ______________ Date ____________

Exploration Lab

FIELD ACTIVITY

Life in a Pine Cone

Large numbers and many varieties of arthropods (insects, spiders, mites, centipedes, millipedes) can live in the structure of a single pine cone. The pine cones afford sanctuary for arthropod inhabitants and visitors to feed, seek mates, lay eggs, hide, or escape danger. Many of the arthropods on or inside cones probably gain access by moving from the soil, up the tree trunk, and out the branches to the cones. Others fly directly to the tree and cones. Entire life histories of some species are completed within cones before and after they fall to the ground and decompose.

The types and numbers of arthropods recovered from pine cones vary with pine species and weather conditions. Most species range in size from about 25 mm down to the size of the period at the end of this sentence.

To separate small arthropods from pine cones, a *Berlese funnel* can be used. This apparatus consists of a large funnel, the mouth of which may be covered with fine wire mesh. The funnel is set into a container. When material that contains small arthropods is placed in the funnel, the organisms collect in the container below while the rest of the material remains in the top of the funnel.

In this lab, you will collect pine cones from trees and from leaf litter and use a Berlese funnel to extract arthropods from the cones. You will identify and compare the arthropods from the two different pine cone microhabitats.

OBJECTIVES

Collect arthropods that live in two different pine cone microhabitats.

Identify the arthropods that live in the pine cone microhabitats.

Compare the diversity of arthropods living in the the two pine cone microhabitats.

MATERIALS

- beakers, small (2)
- Berlese funnel
- cloths, white, about 30 × 30 cm (2)
- felt tip marker
- field guide to arthropods
- filter paper
- forceps or fine paintbrush
- funnel
- gloves
- hand lens
- jar, large
- jars, small (2)
- lab apron
- lamp with 25 W bulb
- plastic bags (4)
- plastic hoop, cloth covered, for collecting cones
- rubbing alcohol
- safety goggles
- sieve, for collecting insects
- stereomicroscope
- trays (2)
- twist ties (4)

Copyright © by Holt, Rinehart and Winston. All rights reserved.

Name ______________________ Class ____________ Date ____________

Procedure

PART 1: COLLECTING AND PRESERVING SPECIMENS

1. Collect at least five pine cones from leaf litter on the ground and five open pine cones from trees. Use a plastic hoop covered with cloth to dislodge the cones from trees. Place the leaf-litter pine cones and the tree pine cones in separate plastic bags. Use a meter to label each bag with the location from which the cones were taken.
2. Use an insect sieve to collect small insects and other small arthropods from the cones. As shown in **Figure 1,** the sieve has a rigid wire ring top, cloth sides, and a wire mesh bottom. Place the sieve over a white cloth, place collected pine cones from one of your bags on the wire mesh, and shake.
3. When you finish shaking the cones, place the white cloth containing the materials you have dislodged in a plastic bag, and tie it securely. Mark the bag with the location (leaf litter or trees) from which the materials were obtained.
4. Use forceps or a fine paintbrush to collect tiny arthropods remaining inside the cones. Carefully pick out or brush the animals and place them in the plastic bag with the other materials you collected in step 3. Tie the bag securely.

FIGURE 1 INSECT SIEVE

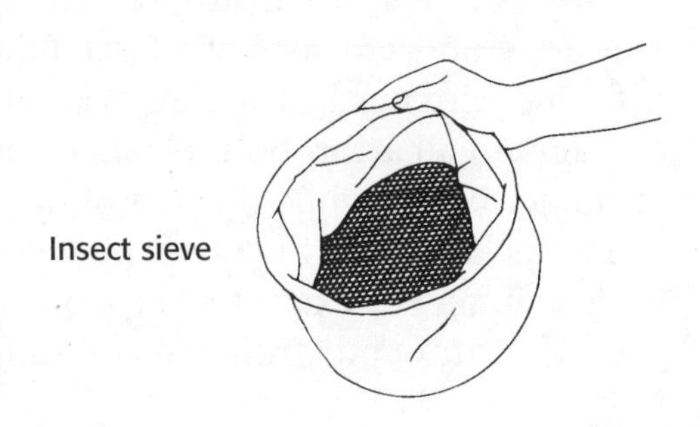

5. Return the cones to their bag.
6. Repeat steps 2-5, using the cones from the other location.
7. Carry your bags of cones and bags of collected materials back to the lab.
8. In the lab, place a small jar inside a large jar, as shown in **Figure 2.** Place the jars under a Berlese funnel. Then place the funnel under a lamp with a 25 W light bulb.
9. Empty one of your bags of cones into the funnel. Turn on the lamp. **CAUTION: Be sure the lamp is at least 5 cm from the cones.** The heat from the bulb will dry the material in the funnel from the top down. As the material warms and dries, the organisms present will move to lower levels and fall through the neck of the funnel into the small jar.

FIGURE 2 APPARATUS FOR DRYING MATERIALS

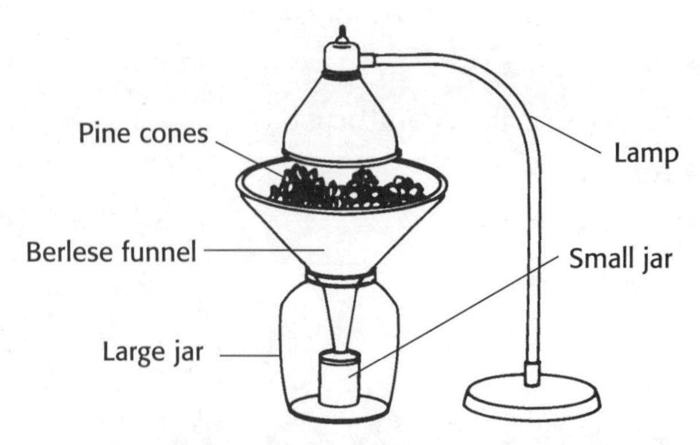

Copyright © by Holt, Rinehart and Winston. All rights reserved.

Name ______________________ Class ______________ Date ____________

Life in a Pine Cone *continued*

10. Put on safety goggles, gloves, and a lab apron.

11. Fill a small beaker $\frac{1}{4}$ full with rubbing alcohol. **CAUTION: Rubbing alcohol is poisonous if swallowed. Avoid contact with eyes.** Transfer the material you collected in the small jar in step 9 to the beaker with rubbing alcohol. The alcohol will preserve the organisms.

12. Add to the alcohol all the other materials you collected (in steps 3 and 4) from the same cones.

13. Separate the materials from the rubbing alcohol by pouring the mixture into a funnel lined with filter paper held over a small jar. Gently transfer the preserved organisms remaining on the filter paper onto a tray. Label the tray with the location (leaf litter or trees) from which the organisms were obtained.

14. Repeat steps 9-13, using your other bag of cones.

PART 2: EXAMINING AND IDENTIFYING SPECIMENS

15. Each of the locations from which you collected cones is a separate microhabitat for arthropods. Examine under a stereomicroscope the preserved arthropods from one of the microhabitats.

16. Group together organisms that are alike, and record them as taxa, or discrete groups of organisms. You do not have to know what each organism is, only that each group has identical individuals.

17. Count the number of individuals in each taxon, and record your tally in the appropriate column in **Table 1.**

TABLE 1 ARTHROPODS COLLECTED FROM TWO MICROHABITATS

Habitat 1 Cones in leaf litter			Habitat 2 Cones on trees		
Taxon	Number of individuals	Name	Taxon	Number of individuals	Name
1	**Sample data: 96**		1	**Sample data: 20**	
2	**4**		2	**3**	
3			3	**30**	
4			4	**37**	
5			5	**10**	

18. Repeat steps 15-17, using the arthropods from the other microhabitat.

19. Use the dichotomous key in **Figure 3** to identify the arthropods you grouped in **Table 1.** You may also want to consult a field guide. In **Table 1,** write the names of the arthropods you are able to identify.

Copyright © by Holt, Rinehart and Winston. All rights reserved.

Name ______________________ Class ______________ Date ______________

20. Dispose of your materials according to your teacher's instructions.

21. Clean up your work area, and wash your hands before leaving the lab.

FIGURE 3 DICHOTOMOUS KEY FOR IDENTIFYING ARTHROPODS

1.	a.	Six legs	Go to 2
	b.	Eight or more legs	Go to 3
2.	a.	No wings	Go to 9
	b.	Wings	Flying insect (Insecta)
3.	a.	Eight legs	Go to 4
	b.	More than eight legs	Go to 7
4.	a.	No cephalothorax	Go to 5
	b.	Cephalothorax present	Go to 6
5.	a.	Long legs	Harvestman (Arachnida)
	b.	Short legs	Mite (Arachnida)
6.	a.	Pincers	Pseudoscorpion (Arachnida)
	b.	No pincers	Spider (Arachnida)
7.	a.	Oval body	Pill bug (Crustacea)
	b.	Long, thin body	Go to 8
8.	a.	One pair of legs per segment	Centipede (Chilipoda)
	b.	Two pairs of legs per segment	Millipede (Diploda)
9.	a.	Chewing mouthparts	Go to 10
	b.	Sucking mouthparts	Go to Scales, Aphid, Leaf hopper, or Cicada (Insecta)
10.	a.	No appendages on abdomen	Go to 11
	b.	Appendages on abdomen	Go to 14
11.	a.	Narrow connection between thorax and abdomen	Ant
	b.	Thick connection between thorax and abdomen	Go to 12
12.	a.	Short antenna	Termite (Insecta)
	b.	Long antenna	Go to 13
13.	a.	All legs the same length	Plant lice (Insecta)
	b.	Longer third pair of legs	Grasshopper, cricket (Insecta)
14.	a.	Hairlike tail	Silverfish (Insecta)
	b.	Forked tail	Springtail (Insecta)

Copyright © by Holt, Rinehart and Winston. All rights reserved.

Name ______________________ Class ______________ Date ____________

Analysis

1. Organizing Data The *diversity index* of a habitat is a number derived from the following formula. In the space provided, use this formula to calculate the diversity index of each microhabitat.

$$\text{Diversity} = \frac{N(N-1)}{\Sigma n(n-1)}$$

Where: N = total number of individuals of all species
n = number of individuals of a species
Σ = the sum of all individuals sampled, as n (n − 1)
Diversity = the number of species in a defined area

Sample calculations based on sample data:

Habitat 1: Cones in leaf litter

$$\frac{100(100-1)}{96(96-1) + 4(4-1)} = \frac{9900}{9132} = 1.08, \text{ the diversity index}$$

Habitat 2: Cones in trees

$$\frac{100(100-1)}{20(20-1) + 3(3-1) + 30(30-1) + 37(37-1) + 10(10-1)} = \frac{9900}{2678} = 3.70, \text{ the diversity index}$$

2. Analyzing Data Compare the diversity of arthropods in the two different pine cone microhabitats.

Answers will vary. For the sample data, the diversity of arthropods in cones found in leaf litter is much less than that found in cones on trees.

Copyright © by Holt, Rinehart and Winston. All rights reserved.

Name ____________ Class ____________ Date ____________

3. **Classifying** Which arthropods did you identify? To which classes do they belong?

   Answers will vary. Encourage students to be as specific as they can, using their data to answer this question.

Conclusions

1. **Drawing Conclusions** How would you account for the difference in diversity of arthropods in the two microhabitats?

   Answers will vary but may include that materials that make up the environment in trees are more varied than those in leaf litter; therefore, trees can support a greater variety of arthropods.

Extensions

1. **Research and Communications** Preserve specimens collected from pine cones permanently for later identification by putting them into a jar with a 50–70% alcohol solution. Find out how to ship the specimens to a systematist for identification. If you ship them, be sure to keep one of each species for future reference.
2. **Designing Experiments** Design a study of organisms in another type of microhabitat that can be carried out in an environment near you. For example, you might collect small arthropods from trees other than pines. You might study the variety of organisms in an open field or along the shore of a lake or stream.

Copyright © by Holt, Rinehart and Winston. All rights reserved.

Skills Practice Lab

RESEARCH

Examining Owl Pellets

Teacher Notes

TIME REQUIRED Two 45-minute periods, one to dissect the owl pellets and another to compile and analyze class data

SKILLS ACQUIRED

Collecting data
Identifying patterns
Inferring
Organizing and analyzing data

RATINGS

Easy ← 1 2 3 4 → Hard

Teacher Prep–2
Student Setup–1
Concept Level–2
Cleanup–2

THE SCIENTIFIC METHOD

Make Observations Students examine pellets of the Northwestern and Southeastern barn owls.

Analyze the Results Analysis questions 2 and 3 require students to analyze their results.

Draw Conclusions Conclusions questions 1, 4, and 5 ask students to draw conclusions from their data.

MATERIALS

Materials for this lab can be purchased from WARD'S. See *the Master Materials List* for ordering instructions.

SAFETY CAUTIONS

- Discuss all safety symbols and caution statements with students.
- Use owl pellets that have been sterilized or fumigated.
- Carefully instruct students on the correct handling of all dissecting instruments.
- Remind students to wash their hands after handling owl pellets.

DISPOSAL

Owl pellets and their contents can be safely disposed of in the trash.

Copyright © by Holt, Rinehart and Winston. All rights reserved.

TIPS AND TRICKS

Preparation

This lab works best in groups of two students.

Owl pellets may be found in roosting and feeding areas, old farm buildings, woodlands, and parks. It is easier to buy pellets that are already dried and sterilized or fumigated than to collect your own from nature. Purchased pellets are accompanied by information and identification guides. If you collect pellets, dry them and fumigate them in polyethylene bags with naphthalene to destroy insect eggs. Soaking the pellets in water for about 2 hours before the lab softens the mucilage and makes dissection easier.

Procedure

Reassure students that, while it may sound unpleasant to dissect something that owls cough up, owl pellets are nothing more than animal bones and tissues 'cleaned' by stomach enzymes. Emphasize that owl pellets are not fecal matter.

Pass out to students owl pellets for dissection. Students must first remove the pellets from the aluminum foil. Then, have students label a sheet of paper with their name. Have them use a dissecting needle to loosen the hair of the owl pellet. As bones are uncovered, they should be carefully removed and placed on the sheet of paper. If you prefer, students can label a small sheet of paper to hold the bones of each prey item that they extract. Egg cartons or petri dishes may be helpful storage containers to use when students are separating the bones into like piles. Suggest that students first sort bones by shape. Then have them sort the bones by size after the relationship among different kinds of bones has been determined. When students identify prey by size, be sure to tell them that some animals may be immature, so size can be deceiving. After all bones have been removed, identification of the skulls can take place.

Another method of identifying raptor prey is through comparison to materials that have already been identified. Most biologists keep a set of identified skulls, along with hair and feather samples. These can be useful when identifying prey remains. This process can speed up the identification of large numbers of similar items by eliminating the need for a key once all of the common items have been identified.

Additional Background

Barn-owl (*Tyto alba)* pellets were chosen for this lab because barn owls feed primarily on small mammals and birds, which are swallowed whole. The heads of long-billed birds and large rats are often removed, but most pellets contain the entire skeletons of birds and small rodents on which the owls have fed.

The nine mammalian prey species listed in **Table 1** account for 96–100% of the prey that students will find. Other prey will consist of birds, bats, and arthropods. These are occasional and too diverse to address in detail.

Copyright © by Holt, Rinehart and Winston. All rights reserved.

Name ______________________ Class ______________ Date ______________

Skills Practice Lab

RESEARCH

Examining Owl Pellets

Owls are *raptors*, or birds of prey. They catch their prey, including small birds and rodents, and swallow them whole, since owls do not have teeth for grinding. Enzymatic juices in the owl's digestive system break down the body tissues of the prey but leave the bony materials, hair, and feathers undigested. Depending on the prey eaten, the undigested portions may include beaks, claws, scales, or insect exoskeletons. These materials have little nutritional value and must be eliminated from the owl's body. They form a *lump* called a *pellet*. Pellets begin forming within the digestive tract of the owl as soon as the prey is swallowed. The pellets are then coughed up, or regurgitated, and the owl begins feeding once again.

Scientists take advantage of this adaptation by collecting owl pellets and analyzing their contents. Owl pellets are dried and either fumigated (treated with chemicals) or sterilized so that their contents can be examined safely. Since owls are not selective feeders, the pellets can be used to estimate the diversity of available prey. The contents are also a direct indicator of the animals on which an owl has fed, which is information that is crucial for species management and protection.

In this lab, you will dissect pellets of barn owls from two different habitats—the Northwest and the Southeast United States. Based on your findings, you will compare the diets of these barn owls.

OBJECTIVES

- **Identify** the species of prey eaten by the Northwestern and Southeastern barn owls from the remains found in the owls' pellets.
- **Compute** the total biomass of each prey and the cumulative total biomass, using class data for each type of barn-owl pellet.
- **Compare** the diets of the Northwestern and Southeastern barn owls, using the data.

MATERIALS

- dissecting needle
- dissecting tray
- forceps
- gloves
- lab apron
- metric ruler
- paper, white
- pellet, Northwestern barn owl
- pellet, Southeastern barn owl
- safety goggles
- tape or glue

Copyright © by Holt, Rinehart and Winston. All rights reserved.

Name ______________________________ Class ______________ Date ____________

Procedure

PART 1: DISSECTING AN OWL PELLET

1. Put on safety goggles, gloves, and a lab apron. **CAUTION: Do not touch your hands to your face or mouth during this lab.**
2. Place an owl pellet in a dissecting tray. Remove the pellet from the aluminum foil casing.
3. Use a dissecting needle and forceps to carefully break apart the owl pellet. **CAUTION: Use sharp instruments with extreme care.** Remove the fur and feathers from the bones. *Note: Be careful to avoid damaging small bones while you are pulling the pellet apart.*
4. As the bones are uncovered, use forceps to carefully place them on a sheet of paper. Take care to remove all skulls and bones from the fur mass. You will identify the animals the owl has eaten mainly by the skulls, mandibles (jaws), and teeth.

   Sometimes undigested beetles and pill bugs are found in owl pellets. These small animals find the expelled pellets and use them as a food source and nursery for their eggs and larvae. Therefore, these organisms should not be included as owl prey.
5. Use the diagrams of the bird and mammal skeletons shown in **Figures 1** and **2**, to help you distinguish among different types of bones.

FIGURE 1 GENERALIZED BIRD SKELETON

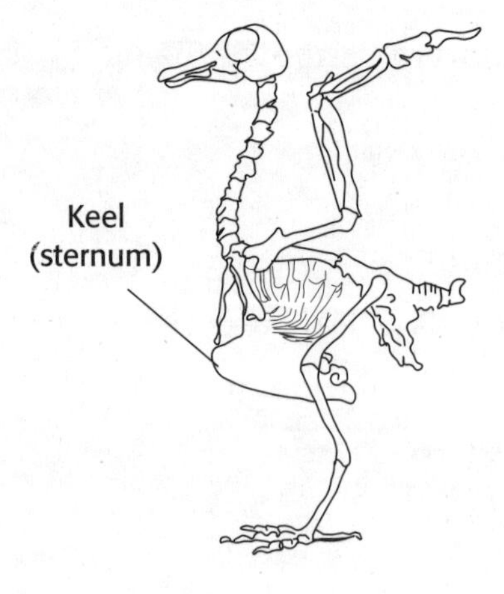

FIGURE 2 GENERALIZED MAMMAL SKELETON

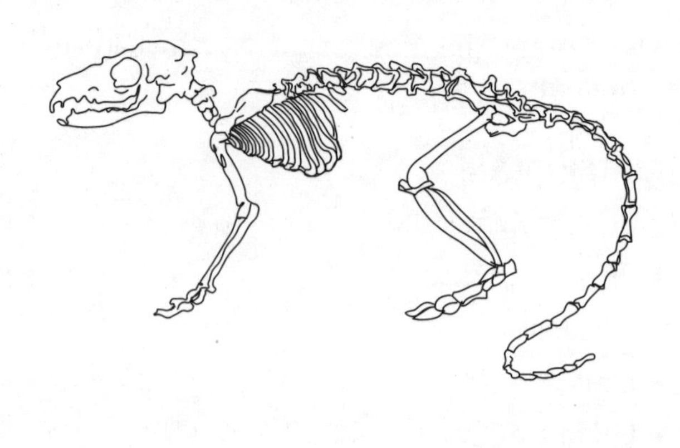

Copyright © by Holt, Rinehart and Winston. All rights reserved.

6. Assemble similar bone parts to see how many prey types are represented in the pellet. Count the number of skulls to determine how many prey were in the pellet. Refer to **Figure 3** to help identify types of skulls.

FIGURE 3 SKULL COMPARISONS

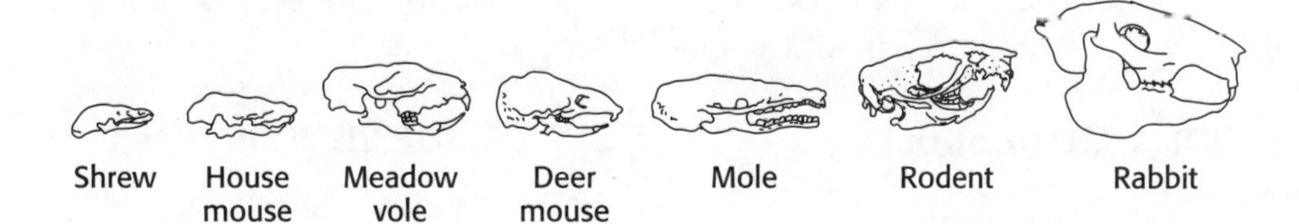

7. Reassemble the skeletons by laying them out on a sheet of paper with appendages arranged and spread out to the side. Glue or tape the assembled skeletons to the paper.
   - How many skeletons were you able to assemble from your owl pellet?

   Answers will vary depending on the owl pellets.

8. Repeat steps 2–7 for the second owl pellet.

PART 2: USING A DICHOTOMOUS KEY TO IDENTIFY PREY

The contents of an owl pellet can be identified with a *dichotomous key*, a tool used to identify an object or an organism. A dichotomous key has a series of statements or questions that compare contrasting characteristics among a group of items or organisms. For example, suppose you want to identify a common U.S. coin by using a dichotomous key. The key might be similar to the one shown in **Figure 4.** Compare the first pair of statements and determine which one best fits the coin you are trying to identify. After you pick one of the paired statements, you will be directed to another paired statement until you reach an answer.

FIGURE 4 DICHOTOMOUS KEY OF COINS

1.	Coin edge smooth	Go to 2
	Coin edge grooved	Go to 3
2.	Coin copper in color	penny
	Coin silver in color	nickel
3.	Picture of Roosevelt on front	dime
	Picture of Washington on front	quarter

Copyright © by Holt, Rinehart and Winston. All rights reserved.

9. Use **Figures 5** and **6,** showing the side and top views of a skull, along with the following information to become familiar with terminology used in a dichotomous key of small mammal skulls.

A gap between the incisors (front teeth) and cheek teeth (molars and premolars) is called a *diastema*. The *infraorbital canal* refers to an opening just below the eye socket. The *zygomatic arch* is a ridge of bone located on the side of the skull making up part of the cheekbone.

FIGURE 5 SIDE VIEW OF SKULL

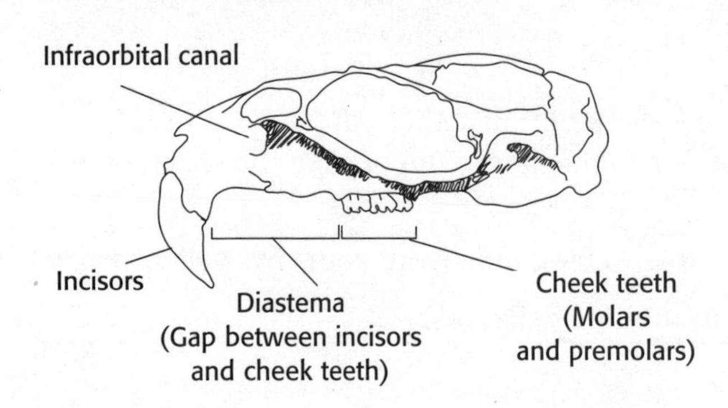

FIGURE 6 TOP VIEW OF SKULL

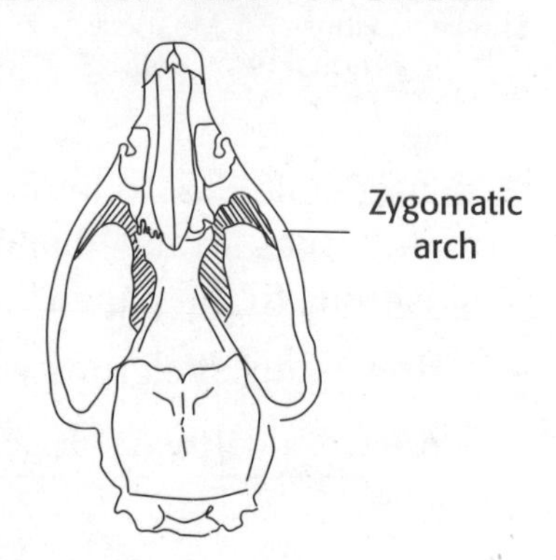

10. Use **Figures 7** and **8** and the following information to help you determine the length of the mandible, or lower jaw.

The bottom view of the skull is actually a view of the roof of the mouth. When determining the posterior edge of the *palate*, look for the position of the *posterior palatine foramina*. The palate ends beyond the posterior palatine foramina.

The skull on the left in **Figure 8** shows the posterior edge of the palate ending even with the last cheek teeth. The right-hand skull shows the palate ending beyond the last cheek teeth.

FIGURE 7 SIDE VIEW OF MANDIBLE (JAW)

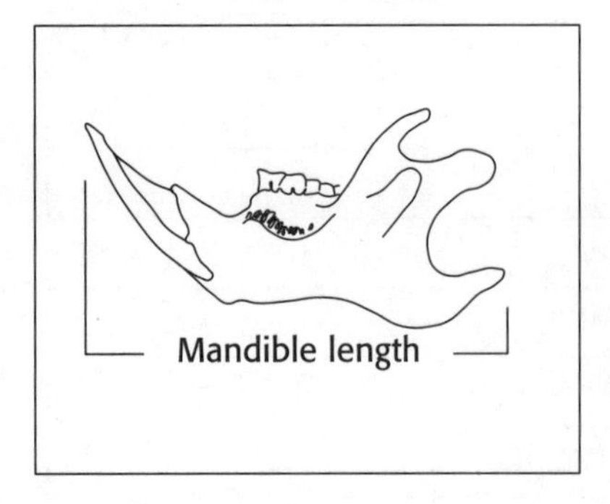

FIGURE 8 BOTTOM VIEW OF SKULLS

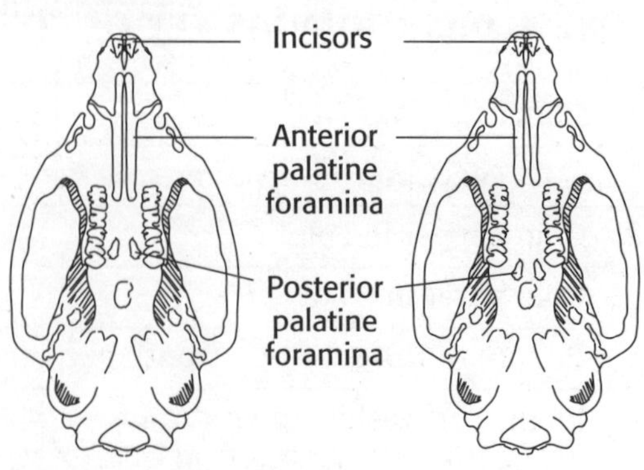

Copyright © by Holt, Rinehart and Winston. All rights reserved.

Name ______________________ Class ______________ Date ____________

Examining Owl Pellets *continued*

11. To determine cheek teeth types or incisor types, refer to **Figures 9** and **10.**

FIGURE 9 CHEEK TEETH TYPES

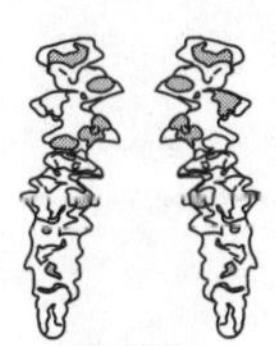
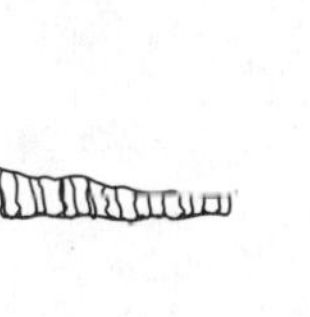
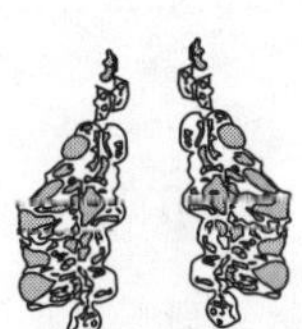
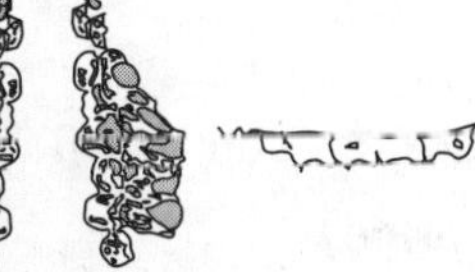
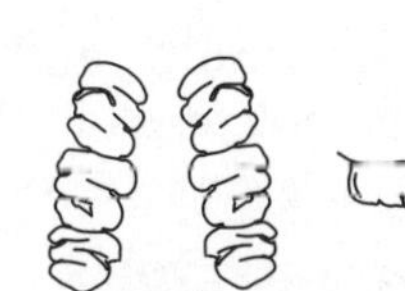

Acute (bottom view) | Acute (side view) | Unicusp (bottom view) | Unicusp (side view) | Lobed s-shaped (bottom view) | Lobed s-shaped (side view)

FIGURE 10 INCISOR TYPES

Grooved incisor (front view) | Smooth incisor (front view) | Notched incisor (front view) | Unnotched incisor (front view)

12. Use the dichotomous key in **Figure 11** and the diagrams in **Figures 5-10** to identify the skulls of small mammals found in your owl pellets. In **Table 1** in the *Number found in pellet* column, record the number of each type of mammal skull found in the Northwestern barn-owl pellet and in the Southeastern barn-owl pellet.

13. Any other animal remains you find in your pellets will be from a bird, a bat, or an insect. List them as 'other prey' in **Table 1.**

- How many different species of animals did you find in each owl pellet?

  Answers will vary depending on the owl pellets.

14. Your teacher will gather class totals for each type of prey in both types of owl pellets. Record these in the *Total number in all pellets* column in **Table 1.**

Copyright © by Holt, Rinehart and Winston. All rights reserved.

Name ______________________ Class ______________ Date ______________

Examining Owl Pellets *continued*

FIGURE 11 DICHOTOMOUS KEY OF SMALL MAMMAL SKULLS

No gap (diastema) between incisors and cheek teeth .... Order: Insectivora
Gap (diastema) between incisors and cheek teeth .... Order: Rodentia

Order Insectivora (moles and shrews)
Zygomatic arch complete; skull flat and broad .... *Scapanus* (mole)
Zygomatic arch not complete; skull not flat and broad .... *Sorex* (shrew)

Order Rodentia (rats, voles, and mice)

1. Infraorbital canal not present .... Go to 2
Infraorbital canal present .... Go to 3

2. Upper incisors distinctly grooved .... *Perognathus* (mouse)
Upper incisors not distinctly grooved .... *Thomomys* (pocket gopher)

3. Skull flat and broad; cheek teeth acutely angled and may appear as one continuous tooth .... *Microtus* (vole)
Skull generally rounded; cheek teeth lobed or rounded and easily distinguished individually .... Go to 4

4. Upper incisors distinctly grooved .... *Reithrodontomys* (harvest mouse)
Upper incisors not distinctly grooved .... Go to 5

5. Posterior edge of palate ending even with or last cheek teeth; cheek teeth not capped with enamel .... *Peromyscus* (deer mouse)
Posterior edge of palate ending beyond last cheek teeth; cheek teeth capped with enamel .... Go to 6

6. Upper incisors notched, mandible length less than 16 mm .... *Mus* (house mouse)
Upper incisors not notched, mandible length greater than 18 mm .... *Rattus* (rat)

15. To compute the total biomass of a prey, multiply the number in the *Total number in all pellets* column by the number listed in the *Prey biomass* column. For example:

If five pocket gophers *(Thomomys)* were recorded as the *Total number in all pellets*, and from the chart the prey biomass is 150 g, then

$$5 \times 150 \text{ g} = 750 \text{ g} = \text{total biomass}$$

Compute totals in this way for all species found, and enter your answers in the *Total biomass* column. Calculate *Cumulative total biomass* by summing all data in the *Total biomass* column.

16. Dispose of your materials according to your teacher's instructions.

17. Clean up your work area, and wash your hands before leaving the lab.

Copyright © by Holt, Rinehart and Winston. All rights reserved.

Name ______________________ Class ______________ Date ____________

Examining Owl Pellets *continued*

TABLE 1 ANALYSIS OF BARN-OWL PREY

Prey	Prey biomass (in grams)	Number found in pellet		Total number in all pellets		Total biomass (in grams)	
		N	S	N	S	N	S
Pocket gopher *Thomomys*	150			7	0	1050	0
Rat *Rattus*	150			2	2	300	300
Vole *Microtus*	40			9	5	360	200
Mice							
Peromyscus	22		**Data**	4	3	88	66
Mus	18		**will**	7	7	126	126
Reithrodontomys	12		**vary.**	4	4	48	48
Perognathus	25			4	0	100	0
Mole *Scapanus*	55			2	0	110	0
Shrew *Sorex*	4			4	1	16	4
Other prey							
bats	7			4	2	28	14
birds	15			5	3	75	45
insects	1			4	3	4	3
Cumulative total biomass (in grams)						2377	860

Key: N = Northwestern barn owl; S = Southeastern barn owl

Sample class data are provided above.

Analysis

1. **Describing Events** How do your data compare with the class's pooled data?

   Answers will vary. Answers should be based on the data students recorded in Table 1.

2. **Analyzing Data** If a barn owl needs 120 g of food per day, how many *Sorex* would it need to capture? How many *Microtus?*

   It would need to capture 30 *Sorex* or 3 *Microtus*.

Copyright © by Holt, Rinehart and Winston. All rights reserved.

Name ______________________ Class ______________ Date ____________

3. **Analyzing Data** Compare the diets of the Northwestern and Southeastern barn owls by comparing the total biomass of each prey to the cumulative total biomass for each type of barn owl.

Answers should be based on the data recorded in Table 1, but in general, pocket gophers make up the largest portion of the Northwestern barn owl's diet. Rats make up the largest portion of the Southeastern barn owl's diet.

Conclusions

1. **Drawing Conclusions** In what way might the formation of owl pellets increase an owl's chances of survival in an ecosystem?

Owl-pellet formation allows owls to be less selective of the kinds of prey on which they can feed. Also, the ability to ingest prey whole speeds up feeding, thereby decreasing exposure to other predators and scavengers.

2. **Making Predictions** How would a sudden decrease in the *Sorex* population affect a barn-owl population? What effect would a decrease in the *Microtus* population have?

Because of its low biomass, *Sorex* does not contribute significantly to a barn owl's diet so, the barn owl population would not be seriously affected. Because of its relatively high biomass, a sudden decrease in *Microtus* would have a detrimental effect on a barn-owl population.

3. **Drawing Conclusions** Assume an owl eats one hundred 1-gram insects and one 100-gram rat. Did the insects or rat contribute more to the owl's diet? Hint: How does foraging time affect this outcome?

The rat contributed more because the owl used less time and energy to capture the rat than it did to capture the insects.

Extension

Research and Communications Predators such as owls are subjected to poisons that have been concentrated through the food chain. Some species of owls are considered endangered, threatened by pesticides and heavy metals that contaminate the foods eaten by their prey. Research this topic and explain how a grain crop contaminated with a compound containing lead can affect an owl population.

Copyright © by Holt, Rinehart and Winston. All rights reserved.

Answer Key

Directed Reading

SECTION: HOW ORGANISMS INTERACT IN COMMUNITIES

1. Coevolution is the back-and-forth evolutionary adjustments between interacting members of an ecosystem. Secondary compounds are defensive chemicals that plants have developed to protect themselves from their predators.
2. Predation is the act of one organism killing another for food. Parasitism is a special case of predation in which one organism, the parasite, lives on or in another organism, the host.
3. coevolution
4. parasitism
5. mutualism
6. commensalism
7. symbiosis
8. mutualism
9. commensalism

SECTION: HOW COMPETITION SHAPES COMMUNITIES

1. c
2. a
3. b
4. d
5. h
6. e
7. i
8. f
9. g
10. *Chthamalus* lives attached to rocks in shallow water and is exposed to air by receding tides. *Semibalanus* lives lower down on the same rocks and is rarely exposed to air.
11. *Chthamalus* was able to live at the lower depth, but *Semibalanus* was unable to live in shallow water.
12. *Semibalanus* would crowd *Chthamalus* off the rocks.
13. *Semibalanus* apparently lacked the adaptation that enabled *Chthamalus* to survive long exposure to air.
14. bacteria
15. more
16. competitive exclusion
17. coexisted with
18. different niches
19. Because sea stars preyed on mussels, eliminating sea stars caused the mussel population to increase. The mussel population crowded out many other species from the rocks.
20. Biodiversity measures the number of different species in a community and the relative number of individuals of each species.
21. The greater the biodiversity of plants in a plot, the greater is the amount of plant material the plot produces.

SECTION: MAJOR BIOLOGICAL COMMUNITIES

1. e
2. g
3. f
4. c
5. d
6. b
7. a
8. climate
9. temperature
10. increases, decreases
11. biome
12. decrease, increases
13. littoral zone
14. limnetic zone
15. profundal zone
16. oceans
17. plankton

Active Reading

SECTION: HOW ORGANISMS INTERACT IN COMMUNITIES

1. C
2. P, M, C
3. P
4. C
5. M, C
6. d

Copyright © by Holt, Rinehart and Winston. All rights reserved.

SECTION: HOW COMPETITION SHAPES COMMUNITIES

1. the first sentence; that Tilman's experiments illustrated the relationship between biodiversity and productivity
2. Tilman monitored plant growth in 147 experimental plots located in a Minnesota prairie.
3. b

SECTION: MAJOR BIOLOGICAL COMMUNITIES

1. biome
2. tropical rain forest, desert, savanna, temperate deciduous forest, temperate grassland, taiga, and tundra
3. Temperature and available moisture decrease.
4. d

Vocabulary Review

1. e
2. m
3. h
4. a
5. f
6. b
7. k
8. d
9. g
10. c
11. j
12. i
13. l
14. climate
15. biome
16. littoral zone
17. limnetic zone
18. profundal zone
19. plankton

Science Skills

INTERPRETING MAPS/ INTERPRETING TABLES

1. polar ice
2. tundra
3. taiga
4. mountain zones
5. tropical rain forest
6. temperate deciduous forest
7. temperate grassland
8. savanna
9. desert or semidesert
10. Tundra and desert; the biomes with the least precipitation support minimal vegetation, and the vegetation of these biomes must be able to survive on limited amounts of water.
11. In general, temperature and available moisture decrease as distance from the equator increases.

Concept Mapping

1. biomes
2. symbiosis
3. predation
4. competition
5. mutualism
6. parasitism
7. niche
8. fundamental niche

Critical Thinking

1. a
2. c
3. d
4. b
5. c
6. e
7. d
8. a
9. b
10. c
11. b
12. d
13. a
14. a, l
15. d, j
16. h, b
17. g, k
18. c, i
19. f, e
20. b
21. a
22. d
23. d

Test Prep Pretest

1. c
2. b
3. a
4. c
5. b
6. b
7. a
8. a
9. b
10. c
11. parasites
12. secondary compounds
13. temperate deciduous forests
14. fundamental niche
15. coevolution
16. declined or decreased
17. tundra
18. equator
19. temperate grasslands
20. A predator typically catches and kills its prey. A parasite lives on or in the organism it feeds on and usually can feed only if the host organism remains alive.
21. The larvae of the cabbage butterfly have evolved the ability to break down

Copyright © by Holt, Rinehart and Winston. All rights reserved.

the defensive chemicals of the plants of the mustard family and therefore feed on them without harm.

22. Predators can prevent one particular species from crowding out other species. This reduces competition among the prey species. Therefore, more species and relative numbers of those species can live in the ecosystem. The reduced competition resulting from the predation increases the biodiversity.

23. The three species of warblers feed in different parts of the spruce tree and in different niches.

24. They are found in the shallow ocean waters of the coastal zones because nutrients washed from land are usually more abundant there.

25. the principle of competitive exclusion

Quiz

SECTION: HOW ORGANISMS INTERACT IN COMMUNITIES

1. c
2. d
3. a
4. b
5. c
6. c
7. d
8. b
9. e
10. a

SECTION: HOW COMPETITION SHAPES COMMUNITIES

1. d
2. c
3. a
4. b
5. a
6. b
7. d
8. a
9. e
10. c

SECTION: MAJOR BIOLOGICAL COMMUNITIES

1. c
2. d
3. d
4. a
5. b
6. e
7. c
8. d
9. a
10. b

Chapter Test (General)

1. c
2. a
3. d
4. a
5. d
6. b
7. d
8. c
9. b
10. b
11. h
12. e
13. f
14. a
15. i
16. b
17. c
18. d
19. j
20. g

Chapter Test (Advanced)

1. b
2. c
3. b
4. d
5. a
6. c
7. b
8. d
9. a
10. c
11. d
12. d
13. a
14. a
15. b
16. d
17. a
18. f
19. e
20. g
21. b
22. c

23. Both the limnetic zone and the surface of the open sea are the waters closest to the surface. Both contain algae, plankton, and fish. The limnetic zone is only in freshwater habitats and is far from shore. The surface waters of the open sea can be close to or far from shore and are part of the marine environment.

24. Marshes and wetlands are intermediate habitats between the open water and the land.

25. No two species can have the same niche. The principle of competitive exclusion states that if two species are competing for the same resources, the species that uses the resources more efficiently will eventually eliminate the other.

Copyright © by Holt, Rinehart and Winston. All rights reserved.

Lesson Plan

Section: How Organisms Interact in Communities

Pacing

Regular Schedule: **with lab(s):** 5 days **without lab(s):** 3 days
Block Schedule: **with lab(s):** 2 1/2 days **without lab(s):** 1 1/2 days

Objectives

1. Describe coevolution.
2. Predict how coevolution can affect interactions between species.
3. Identify the distinguishing features of symbiotic relationships.

National Science Education Standards Covered

UNIFYING CONCEPTS AND PROCESSES

UCP1: Systems, order, and organization

UCP2: Evidence, models, and explanation

UCP4: Evolution and equilibrium

SCIENCE AS INQUIRY

SI1: Abilities necessary to do scientific inquiry

SI2: Understandings about scientific inquiry

SCIENCE IN PERSONAL AND SOCIAL PERSPECTIVES

SPSP2: Population growth

LIFE SCIENCE: INTERDEPENDENCE OF ORGANISMS

LSInter3: Organisms both cooperate and compete in ecosystems.

LSInter4: Living organisms have the capacity to produce populations of infinite size, but environments and resources are finite.

LIFE SCIENCE: MATTER, ENERGY, AND ORGANIZATION IN LIVING SYSTEMS

LSMat4: The complexity and organization of organisms accommodates the need for obtaining, transforming, transporting, releasing, and eliminating the matter and energy used to sustain the organism.

LSMat5: The distribution and abundance of organisms and populations in ecosystems are limited by the availability of matter and energy and the ability of the ecosystem to recycle materials.

Copyright © by Holt, Rinehart and Winston. All rights reserved.

KEY
SE = Student Edition
TE = Teacher Edition
CRF = Chapter Resource File

Block 1

CHAPTER OPENER *(45 minutes)*

_ **Quick Review,** SE. Students answer questions covered in previous sections of the textbook as preparation for the chapter content. **(GENERAL)**

_ **Reading Activity,** SE. Students create a list of all the ways that two species in an ecosystem can interact. They create another list of all the types of communities or ecosystems they can think of. **(GENERAL)**

_ **Using the Figure,** TE. Students answer questions about the chapter opener photograph. **(GENERAL)**

_ **Opening Activity**, TE. Students choose two animals that live in the same community and interact. Then they categorize the animal's interactions under the headings Both Benefit, One Benetifs/One Suffers, One Benefits/One Not Affected. **(GENERAL)**

Block 2

FOCUS *(5 minutes)*

_ **Bellringer Transparency.** Use this transparency as students enter the classroom and find their seats. **(GENERAL)**

MOTIVATE *(10 minutes)*

_ **Identifying Preconceptions**, TE. Students read a passage from the textbook and then describe coevolution.

TEACH *(30 minutes)*

_ **Teaching Transparency, Section Outline.** Use this transparency to give students a framework for the information in this section. **(GENERAL)**

_ **Directed Reading Worksheet, How Organisms Interact in Communities, CRF.** Students complete the exercises in this worksheet to help them understand the material as they read the section. **(BASIC)**

_ **Inclusion Strategies**, TE. Students create a chart of plants around their home or school that are toxic to humans or pets.

_ **Real World Connection**, TE. Students search the Internet for examples of pest repellant plants.

Copyright © by Holt, Rinehart and Winston. All rights reserved.

HOMEWORK

_ **Active Reading Worksheet, How Organisms Interact in Communities, CRF.** Students read a passage related to the section topic and answer questions. **(GENERAL)**

Block 3

TEACH *(25 minutes)*

_ **Data Lab,** Predicting How Predation Would Affect a Plant Species, SE. Students study the effects of grazing on plants. **(GENERAL)**

_ **Datasheets for In-Text Labs, Predicting How Predation Would Affect a Plant Species, CRF.**

_ **Real Life**, SE. Students research effective treatments for the rash caused by poison ivy. **(GENERAL)**

CLOSE *(20 minutes)*

_ **Reteaching,** TE. Students make a chart to summarize the different types of species interactions. **(BASIC)**

_ **Quiz,** TE. Students answer questions that review the section material. **(GENERAL)**

_ **Quiz, CRF.** This quiz consists of ten multiple choice and matching questions that review the section's main concepts. **(BASIC) Also in Spanish.**

HOMEWORK

_ **Section Review,** SE. Assign questions 1–5 for review, homework, or quiz. **(GENERAL)**

Optional Blocks

LAB *(90 minutes)*

_ **Exploration Lab, Life in a Pine Cone, CRF.** Students collect pine cones from trees and from leaf litter and use a Berlese funnel to extract arthropods from the cones. They then identify and compare the arthropods from two different pine cone microhabitats. **(GENERAL)**

Other Resource Options

_ **Internet Connect.** Students can research Internet sources about Symbiosis with SciLinks Code HX4171.

_ **go.hrw.com.** For worksheets, videos, and other teaching aids related to this chapter, visit the HRW Web site and type in the keyword HX4 COM.

_ **CNN Science in the News, Video Segment 15 Greening Sudbury.** This video segment is accompanied by a **Critical Thinking Worksheet**.

Copyright © by Holt, Rinehart and Winston. All rights reserved.

_ **CNN Student News.** Find the latest news, lesson plans, and activities related to important scientific events at **cnnstudentnews.com**.

Copyright © by Holt, Rinehart and Winston. All rights reserved.

Lesson Plan

Section: How Competition Shapes Communities

Pacing

Regular Schedule:	**with lab(s):** N/A	**without lab(s):** 2 days
Block Schedule:	**with lab(s):** N/A	**without lab(s):** 1 day

Objectives

1. Describe the role of competition in shaping the nature of communities.
2. Distinguish between fundamental and realized niches.
3. Describe how competition affects an ecosystem.
4. Summarize the importance of biodiversity.

National Science Education Standards Covered

UNIFYING CONCEPTS AND PROCESSES

UCP1: Systems, order, and organization

UCP2: Evidence, models, and explanation

UCP4: Evolution and equilibrium

SCIENCE AS INQUIRY

SI1: Abilities necessary to do scientific inquiry

SI2: Understandings about scientific inquiry

SCIENCE IN PERSONAL AND SOCIAL PERSPECTIVES

SPSP3: Natural resources

LIFE SCIENCE: INTERDEPENDENCE OF ORGANISMS

LSInter3: Organisms both cooperate and compete in ecosystems.

LSInter4: Living organisms have the capacity to produce populations of infinite size, but environments and resources are finite.

LIFE SCIENCE: MATTER, ENERGY, AND ORGANIZATION IN LIVING SYSTEMS

LSMat4: The complexity and organization of organisms accommodates the need for obtaining, transforming, transporting, releasing, and eliminating the matter and energy used to sustain the organism.

LSMat5: The distribution and abundance of organisms and populations in ecosystems are limited by the availability of matter and energy and the ability of the ecosystem to recycle materials.

Copyright © by Holt, Rinehart and Winston. All rights reserved.

KEY
SE = Student Edition
TE = Teacher Edition
CRF = Chapter Resource File

Block 4

FOCUS *(5 minutes)*

_ **Bellringer Transparency.** Use this transparency as students enter the classroom and find their seats. **(GENERAL)**

MOTIVATE *(10 minutes)*

_ **Discussion Question**, TE. Students discuss why competition is usually most intense between closely related organisms. **(GENERAL)**

TEACH *(30 minutes)*

_ **Teaching Transparency, Section Outline.** Use this transparency to give students a framework for the information in this section. **(GENERAL)**

_ **Teaching Transparency, Warbler Foraging Zone.** Use this transparency to discuss MacArthur's investigation on warblers and how five different warbler species feed on insects in different parts of the same tree. **(GENERAL)**

_ **Data Lab,** Predicting Changes in a Realized Niche, SE. Students analyze a graph showing the prey size and feeding location most frequently selected by one bird species. **(GENERAL)**

_ **Datasheets for In-Text Labs, Predicting Changes in a Realized Niche, CRF.**

HOMEWORK

_ **Directed Reading Worksheet, How Competition Shapes Communities, CRF.** Students complete the exercises in this worksheet to help them understand the material as they read the section. **(BASIC)**

_ **Active Reading Worksheet, How Competition Shapes Communities, CRF.** Students read a passage related to the section topic and answer questions. **(GENERAL)**

Block 5

TEACH *(35 minutes)*

_ **Teaching Transparency, Effects of Competition on an Organism's Niche.** Use this transparency to compare the realized and fundamental niches of *Chthamalus*. Walk students through Connell's experiments. Emphasize that competition can prevent an organism from occupying all of its fundamental niche **(GENERAL)**

_ **Demonstration**, TE. Show photographs of starlings and bluebirds. Describe how starlings have bested bluebirds in the United States and discuss strategies being used to help bluebirds survive. Show students bluebird house plans, for example.

Copyright © by Holt, Rinehart and Winston. All rights reserved.

_ **Group Activity**, Investigating Competition, TE. Students work in small groups to investigate an example of competition. **(ADVANCED)**

_ **Integrating Physics and Chemistry**, TE. Students analyze, review, and critique the scientific explanations of MacArthur, Connell, Gause, Paine, and Tilman about competition.

CLOSE *(5 minutes)*

_ **Quiz,** TE. Students answer questions that review the section material. **(GENERAL)**

HOMEWORK

_ **Alternative Assessment**, Ecological Versus Economic Competition, TE. Students write a short report constrating competition between organisms in a natural community with competition between businesses in a human community. **(ADVANCED)**

_ **Quiz, CRF.** This quiz consists of ten multiple choice and matching questions that review the section's main concepts. **(BASIC) Also in Spanish.**

_ **Section Review,** SE. Assign questions 1–6 for review, homework, or quiz. **(GENERAL)**

Other Resource Options

_ **Career,** Naturalist, TE. Lead a discussion on what a naturalist does and give some background about the field.

_ **go.hrw.com.** For worksheets, videos, and other teaching aids related to this chapter, visit the HRW Web site and type in the keyword HX4 COM.

_ **CNN Science in the News, Video Segment 15 Greening Sudbury.** This video segment is accompanied by a **Critical Thinking Worksheet**.

_ **CNN Student News.** Find the latest news, lesson plans, and activities related to important scientific events at **cnnstudentnews.com**.

Copyright © by Holt, Rinehart and Winston. All rights reserved.

Lesson Plan

Section: Major Biological Communities

Pacing

Regular Schedule: **with lab(s):** 6 days **without lab(s):** 2 days

Block Schedule: **with lab(s):** 3 days **without lab(s):** 1 day

Objectives

1. Recognize the role of climate in determining the nature of a biological community.
2. Describe how elevation and latitude affect the distribution of biomes.
3. Summarize the key features of the Earth's major biomes.
3. Compare and contrast the major freshwater and marine habitats.

National Science Education Standards Covered

UNIFYING CONCEPTS AND PROCESSES

UCP1: Systems, order, and organization

UCP2: Evidence, models, and explanation

UCP4: Evolution and equilibrium

SCIENCE AS INQUIRY

SI1: Abilities necessary to do scientific inquiry

SI2: Understandings about scientific inquiry

SCIENCE IN PERSONAL AND SOCIAL PERSPECTIVES

SPSP2: Population growth

SPSP3: Natural resources

LIFE SCIENCE: MATTER, ENERGY, AND ORGANIZATION IN LIVING SYSTEMS

LSMat5: The distribution and abundance of organisms and populations in ecosystems are limited by the availability of matter and energy and the ability of the ecosystem to recycle materials.

EARTH SCIENCE

ES2: Geochemical cycles

Copyright © by Holt, Rinehart and Winston. All rights reserved.

KEY
SE = Student Edition   **TE** = Teacher Edition
CRF = Chapter Resource File

Block 6

FOCUS *(5 minutes)*

_ **Bellringer Transparency.** Use this transparency as students enter the classroom and find their seats. **(GENERAL)**

MOTIVATE *(10 minutes)*

_ **Activity**, TE. Students develop graphs of year-round temperatures and precipitation in the local area. **(ADVANCED)**

TEACH *(30 minutes)*

_ **Teaching Transparency, Section Outline.** Use this transparency to give students a framework for the information in this section. **(GENERAL)**

_ **Teaching Transparency, Elements of Climate.** Use this transparency to discuss the different types of ecosystems that occur under particular temperature and moisture conditions. **(GENERAL)**

_ **Teaching Transparency, Earth's Biomes.** Use this transparency to discuss the global distribution of seven biomes. Have students make a table showing the types and locations of the biomes of each continent. **(GENERAL)**

_ **Inclusion Strategies**, TE. Students create postcards to send to the class from a biome they have visited. **(GENERAL)**

_ **Teaching Tip**, Latitude and Longitude, TE. Have students complete a graphic organizer that shows how latitude affects climate while longitude is irrelevant to climate. A sample graphic organizer is provided in the TE. **(GENERAL)**

HOMEWORK

_ **Directed Reading Worksheet, Major Biological Communities, CRF.** Students complete the exercises in this worksheet to help them understand the material as they read the section. **(BASIC)**

_ **Active Reading Worksheet, Major Biological Communities, CRF.** Students read a passage related to the section topic and answer questions. **(GENERAL)**

Block 7

TEACH *(35 minutes)*

_ **Group Activity**, Biome Collages, TE. Students collect images and create a collage of each biome. **(GENERAL)**

Copyright © by Holt, Rinehart and Winston. All rights reserved.

_ **Quick Lab,** Investigating Factors That Influence the Cooling of Earth's Surface, SE. Students investigate how the amount of water in an environment affects the rate at which that environment cools. **(GENERAL)**

_ **Datasheets for In-Text Labs, Investigating Factors That Influence the Cooling of Earth's Surface, CRF.**

_ **Integrating Physics and Chemistry**, TE. Ask students whether or not the solubility of gases goes up or down with increasing solvent temperature. **(GENERAL)**

_ **Teaching Transparency, Three Lake Zones.** Use this transparency to describe the littoral, limnetic, and profundal zones of a pond or lake and the characteristic organisms found in each zone. **(GENERAL)**

CLOSE *(10 minutes)*

_ **Reteaching,** TE. Students identify and discuss pictures of different biomes. **(BASIC)**

_ **Quiz,** TE. Students answer questions that review the section material. **(GENERAL)**

HOMEWORK

_ **Quiz, CRF.** This quiz consists of ten multiple choice and matching questions that review the section's main concepts. **(BASIC) Also in Spanish.**

_ **Section Review,** SE. Assign questions 1–5 for review, homework, or quiz. **(GENERAL)**

_ **Science Skills Worksheet, CRF.** Students interpret a map and a table containing information about different biomes. **(GENERAL)**

_ **Modified Worksheet, One-Stop Planner.** This worksheet has been specially modified to reach struggling students. **(BASIC)**

_ **Critical Thinking Worksheet, CRF.** Students answer analogy-based questions that review the section's main concepts and vocabulary. **(ADVANCED)**

Optional Blocks

LABS *(180 minutes)*

_ **Skills Practice Lab,** Observing How Brine Shrimp Select a Habitat, SE. Students investigate habitat selection by brine shrimp and determine which environmental conditions they prefer. **(GENERAL)**

_ **Datasheets for In-Text Labs, Observing How Brine Shrimp Select a Habitat, CRF.**

_ **Skills Practice Lab, Examining Owl Pellets, CRF.** Students dissect pellets of barn owls from two different habitats—the Northwest and the Southeast United States. Based on the findings, students then compare the diets of these barn owls. **(GENERAL)**

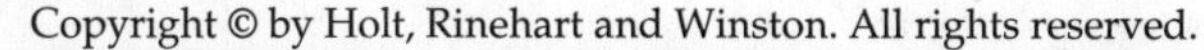
Copyright © by Holt, Rinehart and Winston. All rights reserved.

Other Resource Options

_ **Activity,** Viewing Freshwater Communities, SE. Students collect samples from different parts of a pond, view the samples in aquaria, and make wet mounts. **(GENERAL)**

_ **Internet Connect.** Students can research Internet sources about Biomes with SciLinks Code HX4023.

_ **Internet Connect.** Students can research Internet sources about Estuaries with SciLinks Code HX4073.

_ **Internet Connect.** Students can research Internet sources about Adaptation with SciLinks Code HX4002.

_ **go.hrw.com.** For worksheets, videos, and other teaching aids related to this chapter, visit the HRW Web site and type in the keyword HX4 COM.

_ **CNN Science in the News, Video Segment 15 Greening Sudbury.** This video segment is accompanied by a **Critical Thinking Worksheet**.

_ **CNN Student News.** Find the latest news, lesson plans, and activities related to important scientific events at **cnnstudentnews.com**.

Copyright © by Holt, Rinehart and Winston. All rights reserved.

Lesson Plan

End-of-Chapter Review and Assessment

Pacing

Regular Schedule: 2 days

Block Schedule: 1 day

KEY	
SE = Student Edition	TE = Teacher Edition
CRF = Chapter Resource File	

Block 8

REVIEW *(45 minutes)*

_ **Study Zone,** SE. Use the Study Zone to review the Key Concepts and Key Terms of the chapter and prepare students for the Performance Zone questions. **(GENERAL)**

_ **Performance Zone,** SE. Assign questions to review the material for this chapter. Use the assignment guide to customize review for sections covered. **(GENERAL)**

_ **Teaching Transparency, Concept Mapping.** Use this transparency to review the concept map for this chapter. **(GENERAL)**

Block 9

ASSESSMENT *(45 minutes)*

_ **Chapter Test, Biological Communities, CRF.** This test contains 20 multiple choice and matching questions keyed to the chapter's objectives. **(GENERAL) Also in Spanish.**

_ **Chapter Test, Biological Communities, CRF.** This test contains 25 questions of various formats, each keyed to the chapter's objectives. **(ADVANCED)**

_ **Modified Chapter Test, One-Stop Planner.** This test has been specially modified to reach struggling students. **(BASIC)**

Other Resource Options

_ **Vocabulary Review Worksheet, CRF.** Use this worksheet to review the chapter vocabulary. **(GENERAL) Also in Spanish.**

_ **Test Prep Pretest, CRF.** Use this pretest to review the main content of the chapter. Each question is keyed to a section objective. **(GENERAL) Also in Spanish.**

_ **Test Item Listing for ExamView® Test Generator, CRF.** Use the Test Item Listing to identify questions to use in a customized homework, quiz, or test.

_ **ExamView® Test Generator, One-Stop Planner.** Create a customized homework, quiz, or test using the HRW Test Generator program.

Copyright © by Holt, Rinehart and Winston. All rights reserved.

Biological Communities

TRUE/FALSE

1. ____ Coevolution is the back-and-forth evolutionary adjustments between the interacting members of an ecosystem.
Answer: True   Difficulty: I   Section: 1   Objective: 1

2. ____ A change in the number of predators or prey in a food web can alter the entire ecosystem in which they live.
Answer: True   Difficulty: I   Section: 1   Objective: 2

3. ____ A long-term relationship in which both participating species benefit is known as parasitism.
Answer: False   Difficulty: I   Section: 1   Objective: 2

4. ____ Predation is an example of a biotic interaction.
Answer: True   Difficulty: I   Section: 1   Objective: 2

5. ____ Plants and the herbivores that eat them have evolved independently of one another.
Answer: False   Difficulty: I   Section: 1   Objective: 2

6. ____ Mutualism is a symbiotic relationship in which only one party benefits.
Answer: False   Difficulty: I   Section: 1   Objective: 3

7. ____ An organism's niche includes its habitat.
Answer: True   Difficulty: I   Section: 2   Objective: 1

8. ____ An organism's niche is the sum of all its interactions in its environment, including interactions with other organisms.
Answer: True   Difficulty: I   Section: 2   Objective: 1

9. ____ The total niche an organism is potentially able to occupy within an ecosystem is its realized niche.
Answer: False   Difficulty: I   Section: 2   Objective: 2

10. ____ When two dissimilar species live together in a close association, they are part of a symbiotic relationship.
Answer: True   Difficulty: I   Section: 2   Objective: 3

11. ____ The competitive exclusion principle states that competition usually results in the establishment of cooperation between two species.
Answer: False   Difficulty: I   Section: 2   Objective: 3

12. ____ When ecologist Robert Paine removed the sea star from the 15-species ecosystem along the Washington coast, the number of species in the ecosystem fell to eight.
Answer: True   Difficulty: I   Section: 2   Objective: 4

13. ____ When two species compete for limited resources, competitive exclusion is sure to take place.
Answer: False   Difficulty: I   Section: 2   Objective: 4

14. ____ Warm air can hold more moisture than cold air can hold.
Answer: True   Difficulty: I   Section: 3   Objective: 1

15. ____ Climate is not a factor in determining the ecosystem types found in the United States.
Answer: False   Difficulty: I   Section: 3   Objective: 1

16. ____ Biomes characterized by high annual rainfalls are located at high elevations.
Answer: False   Difficulty: I   Section: 3   Objective: 2

Copyright © by Holt, Rinehart and Winston. All rights reserved.

17. ____ On land, there are 10 major types of ecosystems, which are called biomes.
Answer: False Difficulty: I Section: 3 Objective: 3

18. ____ Permafrost is a characteristic of the taiga.
Answer: False Difficulty: I Section: 3 Objective: 3

19. ____ Deserts are most extensive in the interiors of continents.
Answer: True Difficulty: I Section: 3 Objective: 3

20. ____ Tropical rain forests have the most fertile soil on Earth.
Answer: False Difficulty: I Section: 3 Objective: 3

21. ____ Some animals of the deciduous forest hibernate during the winter.
Answer: True Difficulty: I Section: 3 Objective: 4

22. ____ A biome in which pine forests predominate is the temperate deciduous forest.
Answer: False Difficulty: I Section: 3 Objective: 4

23. ____ Caribou are typical large mammal inhabitants of the tundra.
Answer: True Difficulty: I Section: 3 Objective: 4

24. ____ There are three major types of ecosystems found in the ocean.
Answer: True Difficulty: I Section: 3 Objective: 5

25. ____ All algae are photosynthetic.
Answer: False Difficulty: I Section: 3 Objective: 5

26. ____ Freshwater habitats are independent of terrestrial habitats.
Answer: False Difficulty: I Section: 3 Objective: 5

MULTIPLE CHOICE

27. The process by which species evolve in response to other living members of their ecosystem is called
a. compromise.
b. parasitism.
c. coevolution.
d. ecology.
Answer: C Difficulty: I Section: 1 Objective: 1

28. Over millions of years, plants and their pollinators have
a. coevolved.
b. crossbred.
c. become parasites.
d. become competitive.
Answer: A Difficulty: I Section: 1 Objective: 1

29. The caterpillars of cabbage butterflies are the only insects that can eat plants of the mustard family because they
a. eat these plants only when young and tender.
b. have evolved the ability to break down mustard oils into harmless chemicals.
c. are parasites while in this stage of development.
d. All of the above
Answer: B Difficulty: I Section: 1 Objective: 2

Copyright © by Holt, Rinehart and Winston. All rights reserved.

The diagrams below show different kinds of interactions between species.

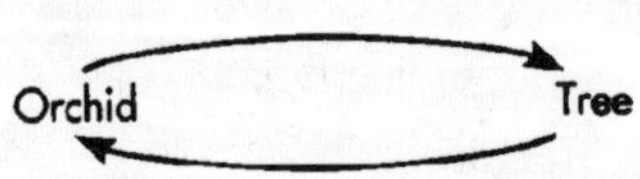

30. Refer to the illustration above. The relationship shown in diagram 1 is
    a. commensalism. c. mutualism.
    b. competition. d. parasitism.

    Answer: C Difficulty: II Section: 1 Objective: 3

31. Refer to the illustration above. The relationship shown in diagram 2 is
    a. commensalism. c. mutualism.
    b. competition. d. parasitism.

    Answer: B Difficulty: II Section: 2 Objective: 1

32. Refer to the illustration above. The relationship shown in diagram 3 is
    a. commensalism. c. mutualism.
    b. competition. d. parasitism.

    Answer: A Difficulty: II Section: 1 Objective: 3

33. Refer to the illustration above. The relationship shown in diagram 4 is
    a. commensalism. c. mutualism.
    b. competition. d. parasitism.

    Answer: D Difficulty: II Section: 1 Objective: 2

34. Parasites
    a. coevolve with their hosts. c. rarely kill their hosts.
    b. are usually smaller than their hosts. d. All of the above

    Answer: D Difficulty: I Section: 1 Objective: 2

35. A tick feeding on a human is an example of
    a. parasitism. c. symbiosis.
    b. mutualism. d. predation.

    Answer: A Difficulty: I Section: 1 Objective: 2

36. Characteristics that enable plants to protect themselves from herbivores include
    a. thorns and prickles. c. chemical defenses.
    b. sticky hairs and tough leaves. d. All of the above

    Answer: D Difficulty: I Section: 1 Objective: 2

37. The relationship between plants and the bees that pollinate them is an example of
    a. commensalism. c. mutualism.
    b. competition. d. parasitism.

    Answer: C Difficulty: I Section: 1 Objective: 3

Copyright © by Holt, Rinehart and Winston. All rights reserved.

38. The relationship between a whale and barnacles growing on its skin is an example of
    a. commensalism.
    b. competition.
    c. mutualism.
    d. parasitism.

    Answer: A   Difficulty: I   Section: 1   Objective: 3

39. The relationship between a clown fish and a sea anemone is known as
    a. parasitism.
    b. competition.
    c. mutualism.
    d. commensalism.

    Answer: D   Difficulty: I   Section: 1   Objective: 3

1	both organisms benefit from the activity of each other
2	one organism benefits and the other organism neither benefits nor suffers harm
3	one organism obtains its nutrients from another; other organism may weaken due to deprivation

40. Refer to the table above. The table represents three types of
    a. competition.
    b. rhythmic patterns.
    c. symbiosis.
    d. secondary succession.

    Answer: C   Difficulty: II   Section: 1   Objective: 3

41. Refer to the table above. Which pair of organisms generally exhibits the type of relationship that corresponds to description 1 in the table?
    a. coyotes and sheep
    b. shrimp and sea cucumbers
    c. parasitic worms and white-tailed deer
    d. ants and aphids

    Answer: D   Difficulty: II   Section: 1   Objective: 3

42. Refer to the table above. The relationship that corresponds to description 2 is known as
    a. parasitism.
    b. commensalism.
    c. mutualism.
    d. predation.

    Answer: B   Difficulty: II   Section: 1   Objective: 3

43. commensalism : one organism ::
    a. parasitism : both organisms
    b. predation : neither organism
    c. competition : both organisms
    d. mutualism : both organisms

    Answer: D   Difficulty: II   Section: 1   Objective: 3

44. Which of the following usually results when members of the same species require the same food and space?
    a. primary succession
    b. competition
    c. secondary succession
    d. interspecific competition

    Answer: B   Difficulty: I   Section: 2   Objective: 1

45. Which of the following would *not* be included in a description of an organism's niche?
    a. its trophic level
    b. the humidity and temperature it prefers
    c. its number of chromosomes
    d. when it reproduces

    Answer: C   Difficulty: I   Section: 2   Objective: 1

46. An organism's niche includes
    a. what it eats.
    b. where it eats.
    c. when it eats.
    d. All of the above

    Answer: D   Difficulty: I   Section: 2   Objective: 1

Copyright © by Holt, Rinehart and Winston. All rights reserved.

47. Most ecosystems tend to be complex because
 a. they are found in all climates.
 b. potential competitors in the ecosystem often occupy slightly different niches.
 c. they all contain a wide variety of producers.
 d. of symbiotic relationships within them.

 Answer: B Difficulty: I Section: 2 Objective: 1

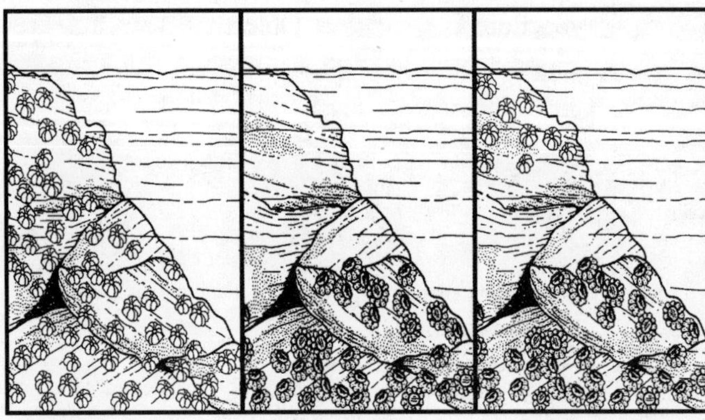

A. The barnacle *Chthamalus stellatus* can live in both shallow and deep water on a rocky coast.

B. The barnacle *Balanus balanoides* prefers to live in deep water.

C. When the two live together, *Chthamalus* is restricted to shallow water.

48. Refer to the illustration above. Because the two species of barnacles attempt to use the same resources, they are
 a. parasitic.
 b. in competition with each other.
 c. mutualistic.
 d. symbiotic.

 Answer: B Difficulty: II Section: 2 Objective: 1
 Answer: B Difficulty: I Section: 2 Objective: 2

49. Refer to the illustration above. Diagram A indicates that the barnacle *Chthamalus stellatus* can live in both shallow and deep water on a rocky coast. This is the barnacle's
 a. competitive niche.
 b. realized niche.
 c. fundamental niche.
 d. exclusive niche.

 Answer: C Difficulty: II Section: 2 Objective: 3

50. Refer to the illustration above. Diagram B indicates that the barnacle *Balanus balanoides* prefers to live in deep water. Deep water is the barnacle's
 a. competitive niche.
 b. realized niche.
 c. fundamental niche.
 d. exclusive niche.

 Answer: C Difficulty: II Section: 2 Objective: 3

51. Refer to the illustration above. Diagram C indicates that when the two barnacles live together, *Chthamalus* is restricted to shallow water. Shallow water is the barnacle's
 a. competitive niche.
 b. realized niche.
 c. fundamental niche.
 d. exclusive niche.

 Answer: B Difficulty: II Section: 2 Objective: 3

52. When two species compete, the niche that each species ultimately occupies is its
 a. competitive niche.
 b. realized niche.
 c. fundamental niche.
 d. exclusive niche.

 Answer: B Difficulty: I Section: 2 Objective: 3

Copyright © by Holt, Rinehart and Winston. All rights reserved.

53. In his experiments with two species of paramecia, G. F. Gause proved that two competitors cannot coexist on the same limited resources. This outcome demonstrated the principle of
    a. competitive exclusion.
    b. secondary succession.
    c. intraspecific competition.
    d. symbiosis.

    Answer: A Difficulty: I Section: 2 Objective: 3

54. If the niches of two organisms overlap,
    a. the organisms may have to compete directly.
    b. the two organisms will always form a symbiotic relationship.
    c. both organisms will disappear from the habitat.
    d. one organism usually migrates to a new habitat.

    Answer: A Difficulty: I Section: 2 Objective: 4

55. Sea stars are fierce competitors of marine organisms such as clams and mussels. An ecologist studying an ocean ecosystem performed an experiment in which the sea stars were removed from the ecosystem. After the removal of the sea stars,
    a. the ecosystem became more diverse.
    b. the size of the ecosystem was reduced.
    c. food webs in the ecosystem became more complex.
    d. the number of species in the ecosystem was reduced.

    Answer: D Difficulty: I Section: 2 Objective: 4

56. Major ecosystems that occur over wide areas of land are called
    a. communities.
    b. habitats.
    c. biomes.
    d. food chains.

    Answer: C Difficulty: I Section: 3 Objective: 1

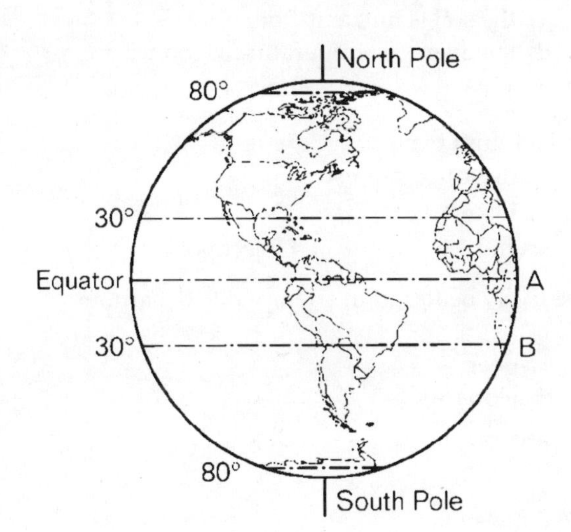

57. Refer to the illustration above. An ecosystem located along latitude A would
    a. have a shorter growing season than an ecosystem on latitude B.
    b. probably contain fewer species than an ecosystem at latitude B.
    c. probably be more diverse than an ecosystem at latitude B.
    d. probably have less rainfall than an ecosystem at latitude B.

    Answer: C Difficulty: II Section: 3 Objective: 2

58. Generally, the closer an ecosystem is to the equator,
    a. the longer its growing season.
    b. the greater its diversity.
    c. the warmer its temperature.
    d. All of the above

    Answer: D Difficulty: I Section: 3 Objective: 2

Copyright © by Holt, Rinehart and Winston. All rights reserved.

59. Generally, the closer an area is to the equator, the greater the diversity in species. Following are the latitudes of four cities. Which city would you predict to have the greatest diversity of species?
    a. Berlin, Germany (latitude: about 52° N)
    b. Montreal, Canada (latitude: 45° N)
    c. Denver, Colorado (latitude: about 40° N)
    d. Brisbane, Australia (latitude: about 28° S)

    Answer: D   Difficulty: I   Section: 3   Objective: 2

60. The biome that makes up most of the central part of the United States is
    a. rain forest.
    b. temperate grassland.
    c. tundra.
    d. deciduous forest.

    Answer: B   Difficulty: I   Section: 3   Objective: 3

61. Tropical ecosystems are more diverse than temperate zone ecosystems because
    a. the growing season in tropical ecosystems never stops.
    b. the climate in tropical ecosystems does not vary much from year to year.
    c. a greater amount of food is produced in tropical ecosystems.
    d. All of the above

    Answer: D   Difficulty: I   Section: 3   Objective: 3

62. Which of the following biomes is characterized by evergreen trees and mammals such as moose, bears, and lynx?
    a. taiga
    b. polar
    c. temperate deciduous forest
    d. tundra

    Answer: A   Difficulty: I   Section: 3   Objective: 4

63. The most likely reason there are few large trees in the tundra is because
    a. they are grazed by bison.
    b. permafrost restricts root development.
    c. the soil is nutrient-poor.
    d. the average temperature is too warm.

    Answer: B   Difficulty: I   Section: 3   Objective: 4

64. Herds of grazing animals are most likely to be found in a
    a. savanna.
    b. tropical rain forest.
    c. deciduous forest.
    d. taiga.

    Answer: A   Difficulty: I   Section: 3   Objective: 4

65. Which of the following animals would most likely be found in a temperate deciduous forest?
    a. monkeys
    b. caribou
    c. deer
    d. leopards

    Answer: C   Difficulty: I   Section: 3   Objective: 4

66. Plankton are
    a. a major formation ingredient of most fossil fuels.
    b. found in the deep-water zone of most lakes and ocean.
    c. the base of most aquatic food webs.
    d. usually in the third and fourth trophic levels of ocean ecosystems.

    Answer: C   Difficulty: I   Section: 3   Objective: 4

67. Organisms with light-producing body parts would most likely be found in
    a. the deep-water zone of lakes.
    b. shallow ocean waters.
    c. open ocean surfaces.
    d. deep ocean waters.

    Answer: D   Difficulty: I   Section: 3   Objective: 4

68. The greatest diversity of life in the ocean is found in
    a. shallow ocean waters.
    b. the ocean surface.
    c. deep ocean waters.
    d. tidal areas.

    Answer: A   Difficulty: I   Section: 3   Objective: 5

Copyright © by Holt, Rinehart and Winston. All rights reserved.

69. Almost all of the Earth's surface water is contained in
 a. ocean ecosystems.
 b. freshwater biomes.
 c. tropical rain forests.
 d. ponds and lakes.
 Answer: A Difficulty: I Section: 3 Objective: 5

70. many fish : shallow-ocean-water habitat ::
 a. nutrients : deep-sea waters
 b. plankton : deep-sea-water habitat
 c. plankton : open-sea surface habitat
 d. animals producing own light : shallow-ocean-water habitat
 Answer: C Difficulty: II Section: 3 Objective: 5

COMPLETION

71. The back-and-forth evolutionary adjustments between interacting members of an ecosystem are called ____________________.
 Answer: coevolution Difficulty: II Section: 1 Objective: 1

72. In a parasitic relationship, the organism that provides benefits to another organism at its own expense is called the ____________________.
 Answer: host Difficulty: I Section: 1 Objective: 2

73. The general term for the biotic relationship in which one organism feeds upon another is ____________________.
 Answer: predation Difficulty: I Section: 1 Objective: 2

74. A symbiotic relationship in which one organism benefits and another is often harmed but not killed is called ____________________.
 Answer: parasitism Difficulty: I Section: 1 Objective: 2

75. When two or more species evolve in response to each other, it is called ____________________.
 Answer: coevolution Difficulty: II Section: 1 Objective: 2

76. The symbiotic relationship in which one organism benefits and the other neither benefits nor suffers harm is called ____________________.
 Answer: commensalism Difficulty: I Section: 1 Objective: 3

77. The term ____________________ is used to describe a close relationship between two dissimilar organisms in which one organism usually benefits.
 Answer: symbiosis Difficulty: I Section: 1 Objective: 3

78. A fish called a cleaner wrasse eats the tiny parasites that cling to and feed upon much larger fish. Therefore, the cleaner wrasse has a(n) ____________________ relationship with the larger fish.
 Answer: mutualistic Difficulty: II Section: 1 Objective: 3

79. A(n) ____________________ describes the habitat, feeding habits, other aspects of an organism's biology, and its interactions with other organisms and the environment.
 Answer: niche Difficulty: I Section: 2 Objective: 1

80. The struggle among organisms for the same limited natural resources is called ____________________.
 Answer: competition Difficulty: I Section: 2 Objective: 1

81. The total niche that an organism is potentially able to use within an ecosystem is called that organism's ____________________ ____________________.
 Answer: fundamental niche Difficulty: I Section: 2 Objective: 2

Copyright © by Holt, Rinehart and Winston. All rights reserved.

82. Local elimination of one competing species is called ____________________
____________________.
Answer: competitive exclusion Difficulty: II Section: 2 Objective: 3

83. The variety of organisms in a community is called ____________________.
Answer: biodiversity Difficulty: I Section: 2 Objective: 4

84. The prevailing weather conditions in any given area is called the area's
____________________.
Answer: climate Difficulty: I Section: 3 Objective: 1

85. A major biological community that occurs over a large area of land is called a(n)
____________________.
Answer: biome Difficulty: I Section: 3 Objective: 2

86. The thick, continually frozen layer of ground found in the northern tundra is called
____________________.
Answer: permafrost Difficulty: I Section: 3 Objective: 3

87. The biome that makes up most of the central part of the continental United States is
____________________ ____________________.
Answer: temperate grasslands Difficulty: I Section: 3 Objective: 3

88. A(n) ____________________ is open, windswept ground that is always frozen.
Answer: tundra Difficulty: I Section: 3 Objective: 3

89. Coniferous trees are predominantly found in the ____________________ biome.
Answer: taiga Difficulty: I Section: 3 Objective: 4

90. Trees that lose their leaves every year are known as ____________________.
Answer: deciduous Difficulty: I Section: 3 Objective: 4

91. Elk and moose may live in the ____________________, areas that are also the primary source of the world's lumber.
Answer: taiga Difficulty: I Section: 3 Objective: 4

92. A biome that is characterized by lush vegetation, abundant rain, and year-round warm temperatures is a(n) ____________________ ____________________
____________________.
Answer: tropical rain forest Difficulty: I Section: 3 Objective: 4

93. A dry grassland known as a(n) ____________________ is the home of elephants, giraffes, and lions, having open, widely spaced trees and seasonal rainfall.
Answer: savanna Difficulty: I Section: 3 Objective: 4

94. Ocean waters that are ____________________ are small in area but contain most of the ocean's diversity.
Answer: shallow Difficulty: I Section: 3 Objective: 4

95. Freshwater marshes and wetlands are often ____________________
____________________ between freshwater and terrestrial habitats.
Answer: intermediate habitats
Difficulty: II Section: 3 Objective: 5

96. Not all freshwater systems are deep enough to include a(n) ____________________ zone.
Answer: profundal Difficulty: II Section: 3 Objective: 5

Copyright © by Holt, Rinehart and Winston. All rights reserved.

PROBLEM

The data in the table shown below were taken during a study of an abandoned agricultural field. Scientists counted the number of different kinds of herbs, shrubs, and trees present in the field 1, 25, and 40 years after it had been abandoned.

	Time after abandonment of agricultural field		
	1 year	**25 years**	**40 years**
Number of herb species	31	30	36
Number of shrub species	0	7	19
Number of tree species	0	14	2
Total number of species	31	51	77

97. Write three conclusions that you can draw from these data.
    Answer:
    The following are some possible conclusions:
    1. The total number of plant species present in the field increased over the 40-year time period.
    2. The plants that grew initially in the field were all herbs.
    3. Over the 40-year time period, the relative proportions of herbs, shrubs, and trees changed. The relative number of herbs decreased, while the relative number of shrubs and trees increased.
    4. The total number of herbs present did not change significantly over the 40-year time period.

    Difficulty: III Section: 2 Objective: 3

98. Make a prediction of the relative numbers of herbs, shrubs, trees, and of the total number of plant species that you would expect to see 100 years after abandonment of the field.
    Answer:
    It is likely that the total number of species present would be even greater 100 years after abandonment. There would probably be relatively fewer herbs, about the same or relatively more shrubs, and relatively more trees.

    Difficulty: III Section: 2 Objective: 3

ESSAY

99. To control its wild rabbit population, the Australian government introduced the viral disease myxomatosis. At first the virus was very deadly to the rabbits but eventually the rabbit population stabilized. Explain what happened to the rabbit population and why the virus became less virulent.
    Answer:
    At first the virus was very deadly, so rabbits that survived tended to be resistant to the virus. The virus also became less deadly because viral infections that kill their hosts too quickly are not able to spread to as many other rabbits. The rabbits and virus coevolved.

    Difficulty: III Section: 1 Objective: 1

Copyright © by Holt, Rinehart and Winston. All rights reserved.

100. Some species of orchids grow high in the trees of tropical forests. The trees provide the orchids with the support to grow and allow them to capture more sunlight than they would on the forest floor. What form of symbiosis is illustrated by this occurrence? Explain your answer.

Answer:

Commensalism is the form of symbiosis illustrated. In commensalism, one organism benefits and the other organism neither benefits nor suffers harm. In this example, the orchids benefit from the presence of the trees, but the trees are not harmed because the orchids neither feed on their tissues nor prevent significant amounts of sunlight from reaching their leaves.

Difficulty: III Section: 1 Objective: 3

101. Describe the type of biotic interaction called competition.

Answer:

All organisms compete for food, water, space, and other resources. One type of competition occurs between members of the same species, and another type of competition occurs between different species.

Difficulty: II Section: 2 Objective: 1

102. Which type of organisms are most likely to survive—those that have a narrow ecological niche or those that have a broad niche? Explain.

Answer:

Organisms having broad niches are more likely to survive because they are not likely to depend on a single food source or a single habitat. If one food source becomes scarce, they can turn to another; or if one habitat is destroyed they can move to another. An organism having a narrow niche may depend totally on a single food source or require a specific habitat. If the food source or habitat is disrupted, the organism may not survive.

Difficulty: III Section: 2 Objective: 4

103. Can two species occupy the same niche? Explain.

Answer:

No two species can have the same niche. The principle of competitive exclusion states that if two species are competing for the same resource, the species that uses the resource more efficiently will eventually eliminate the other.

Difficulty: II Section: 2 Objective: 4

104. Explain and give an example of what is meant by the statement "Climate has an important influence on the type of ecosystem found in an area."

Answer:

The climate of an area refers to the daily atmospheric conditions—the temperature, amount of rainfall, and amount of sunlight in a given area. The physical features of the Earth and amount of solar energy reaching an area influence the climate. Ecosystems vary based on the types of living organisms—plants and animals—that can survive in an area.

Areas receiving large amounts of sunlight and precipitation tend to be warm and moist and support different types of organisms than do colder, dry areas. Areas that are warm and dry, such as parts of southern Arizona, promote the growth of fewer plants than areas with heavy rainfall. The plants that survive, such as cacti, have developed structures that promote water conservation. Areas with mild temperatures and heavier rainfall, such as Virginia and North Carolina, promote the growth of dense forests with tall trees that shed their leaves and consume large amounts of water on a daily basis. (Other examples also are acceptable that establish a link between the type of organisms that can survive and the climate.)

Difficulty: III Section: 3 Objective: 1

Copyright © by Holt, Rinehart and Winston. All rights reserved.

105. Why are plankton important in freshwater ecosystems? Are plankton important in land ecosystems as well? Explain.

Answer:

Plankton, a diverse biological community of microscopic organisms, live near the surface of lakes and ponds. Plankton contain photosynthetic organisms that are the base of aquatic food webs. Plankton are important in land ecosystems because these ecosystems are closely connected to freshwater habitats. Many land animals come to water to feed on aquatic animals that rely on plankton or plankton-eating organisms for food.

Difficulty: II Section: 3 Objective: 4

Copyright © by Holt, Rinehart and Winston. All rights reserved.
9997265955 1 2 3 4 5 6